# Reliability, Maintainability, and Safety for Engineers

# Reliability, Maintainability, and Safety for Engineers

B.S. Dhillon

CRC Press is an imprint of the
Taylor & Francis Group, an **informa** business

First edition published 2020
by CRC Press
6000 Broken Sound Parkway NW, Suite 300, Boca Raton, FL 33487-2742

and by CRC Press
2 Park Square, Milton Park, Abingdon, Oxon, OX14 4RN

First issued in paperback 2021



CRC Press is an imprint of Taylor & Francis Group, an Informa business



**Visit the Taylor & Francis Web site at**
http://www.taylorandfrancis.com

**and the CRC Press Web site at**
http://www.crcpress.com

*Library of Congress Cataloging-in-Publication Data*

ISBN 13: 978-1-03-224191-3 (pbk)
ISBN 13: 978-0-367-35265-3 (hbk)

Typeset in Palatino
by Cenveo® Publisher Services

# *Dedication*

---

*This book is affectionately dedicated*
*to my younger brother*
*Balwant S. Dhillon and his*
*wife Debbi Dhillon.*

# Contents

# Preface

Nowadays, engineering systems/products are an important element of the world's economy, and each year, billions of dollars are spent to develop, manufacture, operate, and maintain various types of engineering systems/products around the globe. Many of these systems/products are highly sophisticated and contain millions of parts. For example, a Boeing Jumbo 747 is made up of approximately 4.5 million parts including fasteners. Needless to say, reliability, maintainability, and safety of systems/products such as this have become more important than ever before. Global competition and other factors are forcing manufacturers to produce highly reliable, maintainable, and safe engineering systems/products.

It means that there is a definite need for the reliability and safety professionals to work closely during design and other phases. To achieve this goal, it is essential that they have an understanding of each other's discipline to a certain degree. At present, to the best of the author's knowledge, there is no book that covers the topics of reliability, maintainability, and safety within its framework. It means, at present to gain knowledge of each other's specialities, these specialists must study various books, reports, or articles on each of the topics in question. This approach is time-consuming and rather difficult because of the specialized nature of the material involved.

Thus, the main objective of this book is to combine these three topics into a single volume and to eliminate the need to consult many different and diverse sources in obtaining basic and up-to-date desired information on the topics. The sources of most of the material presented are given in the Reference section at the end of each chapter. This will be useful to readers if they desire to delve more deeply into a specific area or topic. The book contains a chapter on mathematical concepts and another chapter on the basics of reliability, maintainability, and safety considered useful to understand the contents of subsequent chapters. Furthermore, another chapter is devoted to methods considered useful to analyze the reliability, maintainability, and safety of engineering systems.

The topics covered in the book are treated in such a manner that the reader will require no previous knowledge to understand contents. At appropriate places, the book contains examples along with their solutions, and there are numerous problems at the end of each chapter to test the reader's comprehension in the area.

The book comprises 14 chapters. Chapter 1 presents the need for and historical developments in reliability, maintainability, and safety; engineering systems reliability/maintainability/safety-related facts, figures, and examples; important terms and definitions; and useful sources for obtaining information on reliability, maintainability, and safety. Chapter 2 reviews mathematical concepts considered useful to understand subsequent chapters. Some of the topics covered in the chapter are Boolean algebra laws, probability properties, probability distributions, and useful mathematical definitions.

Chapter 3 presents various introductory aspects of reliability, maintainability, and safety. Chapter 4 presents a number of methods considered useful to analyze engineering systems reliability, maintainability, and safety. These methods are failure modes and effect analysis, fault tree analysis, the Markov method, cause and effect diagram, probability tree analysis, hazards and operability analysis, technique of operations review, job safety analysis, and interface safety analysis. Chapter 5 is devoted to reliability management. Some of the topics covered in the chapter are general management reliability program responsibilities, a procedure for developing reliability goals, reliability management documents and tools, reliability engineering department responsibilities, and pitfalls in reliability program management.

Chapter 6 presents myriad aspects of human and mechanical reliability. Some of the topics covered in the chapter are human error types and causes, human performance effectiveness and stress factors, mean time to human error measure, human reliability analysis methods, mechanical failure modes and causes, safety factors and safety margin, and stress–strength interference theory modeling. Chapter 7 is devoted to reliability testing and growth. Some of the topics covered in the chapter are reliability test classifications, success testing, accelerated life testing, reliability growth program, reliability growth process evaluation approaches, and reliability growth models.

Chapter 8 discusses various important aspects of maintainability management. Some of the topics covered in the chapter are maintainability management functions during the product life cycle, maintainability organization functions, maintainability program plan, and maintainability design reviews. Chapter 9 is devoted to human factors in maintainability. Some of the topics covered in the chapter are typical human behaviors, human sensory capabilities, visual and auditory warnings devices in maintenance activities, and human factors formulas.

Chapter 10 presents various important aspects of maintainability testing and demonstration. Some of the topics covered in the chapter are maintainability testing and demonstration planning and control requirements, maintainability test approaches, maintainability testing methods, and steps for performing maintainability demonstrations and evaluating the results. Chapter 11 is devoted to safety management. Some of the topics covered in the chapter are principles of safety management; functions of safety department, manager, and engineer; steps for developing a safety program plan; product safety management program; and safety performance measures.

Chapter 12 presents various important aspects of safety costing. Some of the topics covered in the chapter are safety cost-related facts, figures, and examples; losses of a company due to an accident involving its product; safety cost estimation methods; and safety cost performance measurement indexes. Chapter 13 is devoted to human factors in safety. Some of the topics covered in the chapter are job stress, worksite analysis program for human factors, useful Occupational Safety and Health Administration ergonomics guidelines, and human factors related safety issues.

Finally, Chapter 14 presents various important aspects of software and robot safety. Some of the topics covered in the chapter are software hazard causing ways and safety classifications, software safety assurance program, software hazard analysis methods, robot safety problems and accident types, robot hazard causes, and robot safeguard approaches.

This book will be useful to many individuals, including design engineers; system engineers; reliability, maintainability, and safety professionals; engineering administrators; graduate and undergraduate students in the area of engineering; researchers and instructors of reliability, maintainability, and safety; and engineers at large.

The author is deeply indebted to many individuals, including family members, colleagues, friends, and students for their invisible inputs. The invisible contributions of my children are also appreciated. Last, but not least, I thank my wife, Rosy, my other half and friend, for typing this entire book and for timely help in proofreading.

**B.S. Dhillon**
*University of Ottawa*

# About the author

**Dr. B.S. Dhillon** is a professor of Engineering Management in the Department of Mechanical Engineering at the University of Ottawa. He has served as a chairman/director of Mechanical Engineering Department/ Engineering Management Program for over 10 years at the same institution. He is the founder of the probability distribution named *Dhillon Distribution/ Law/Model* by statistical researchers in their publications around the world. He has published over 402 (i.e., 244 [70 single authored + 174 co-authored] journal and 158 conference proceedings) articles on reliability engineering, maintainability, safety, engineering management, etc. He is or has been on the editorial boards of 12 international scientific journals. In addition, Professor Dhillon has written 47 books on various aspects of health care, engineering management, design, reliability, safety, and quality published by Wiley (1981), Van Nostrand (1982), Butterworth (1983), Marcel Dekker (1984), Pergamon (1986), etc. His books are being used in over 100 countries, and many of them are translated into languages such as German, Russian, Chinese, Arabic, and Persian (Iranian).

Professor Dhillon has served as general chairman of two international conferences on reliability and quality control held in Los Angeles and Paris in 1987. He has also served as a consultant to various organizations and bodies and has many years of experience in the industrial sector. He has lectured in over 50 countries, including keynote addresses at various international scientific conferences held in North America, Europe, Asia, and Africa. In March 2004, Professor Dhillon was a distinguished speaker at the Conference/Workshop on Surgical Errors (sponsored by White House Health and Safety Committee and Pentagon) held at the Capitol Hill (One Constitution Avenue, Washington, D.C.).

Professor Dhillon attended the University of Wales where he received a B.S. degree in electrical and electronic engineering and an M.S. degree in mechanical engineering. He received his Ph.D. degree in industrial engineering from the University of Windsor.

# *chapter one*

# *Introduction*

## *1.1 Background*

The history of the reliability field may be traced back to the early years of 1930s when probability concepts were applied to electric power generation-associated problems [1, 2]. During World War II, Germans applied the basic reliability concepts to improve reliability of their V1 and V2 rockets. During the period of 1945–1950, the U.S. Department of Defense conducted various studies concerning electronic equipment failure, equipment maintenance, repair cost, etc. As the result of the findings of these studies, it formed an ad hoc committee on reliability, and in 1952, the committee was transformed to a permanent body: The Advisory Group on the Reliability of Electronic Equipment. Additional information on the history of the reliability field is available in Ref. [3].

Although, the precise origin of maintainability as an identifiable discipline is somewhat obscured, but in some ways, the concept goes back to the very beginning of the twentieth century. For example, in 1901, the Army Signal Corps contract for the development of the Wright Brothers' airplane stated that the aircraft should be 'simple to operate and maintain' [4]. In the modern context, the beginning of the discipline of maintainability may be traced to the period between World War II and the early years of 1950s, when various studies carried out by the U.S. Department of Defense and produced startling results [5, 6]. For example, a Navy study reported that during maneuvers, electronic equipment was operative only 30% of the time. Additional information on the history of maintainability is available in Refs. [7, 8].

The history of the safety field may be traced back to the Code of Hammurabi (2000 BC) developed by a Babylonian ruler named Hammurabi. However, in modern times, in 1868, a patent was awarded for the first barrier safeguard in the United States [9]. Twenty-five years later in 1893, the U.S. Congress passed the Railway Safety Act, and in 1912, the Cooperative Safety Congress met in Milwaukee, Wisconsin [9, 10]. Additional information on the history of safety is available in Ref. [11].

## 1.2 Reliability, maintainability, and safety facts, figures, and examples

Some of the facts, figures, and examples, directly or indirectly, concerned with engineering system reliability/maintainability/safety are as follows:

- As per Refs. [12, 13], the number of persons killed because of computer system-related failures was somewhere between 1000 and 3000.
- Each year, the U.S. industry spends about $300 billion on plant maintenance and repair [14].
- A study by the U.S. Nuclear Regulatory Commission reported that around 65% of nuclear system failures involve human error [15].
- In 2002, a study commissioned by the National Institute of Standards and Technology reported that software errors cost the U.S. economy about US $59 billion per year [16].
- A study reported that around 12%–17% of the accidents in the industrial sector using advanced manufacturing technology were related to automated production equipment [17, 18].
- In a typical year, the work accidental deaths by cause in the United States are motor vehicle related: 37.2%, falls: 12.5%, electric current: 3.7%, drowning: 3.2%, fire related: 3.1%, air transport related: 3%, poison (solid, liquid): 2.7%, water transport related: 1.65%, poison (gas, vapor): 1.4%, and others: 31.6% [9, 19].
- In the European Union, approximately 5500 persons are killed due to workplace-related accidents each year [20].
- In 1969, the U.S. Department of Health, Education, and Welfare special committee reported that over a period of ten years, there were around 10,000 medical device-related injuries and 731 resulted in deaths [21, 22].
- As per Ref. [23], some studies carried out in Japan indicate that more than 50% of working accidents with robots can be attributed to faults in the control systems' electronic circuits.
- A study reported that approximately 18% of all aircraft accidents are maintenance related [24, 25].
- A study of safety-related issues concerning onboard fatalities of jet fleets worldwide for the period of 1982–1991 reported that inspection and maintenance were clearly the second most important safety issue, with a total of 1481 onboard fatalities [26, 27].
- A study of over 4400 maintenance-related records concerning a boiling water reactor nuclear power plant covering the period from 1992 to 1994 reported that around 7.5% of all failure records could be attributed to human error related to maintenance tasks/activities [28, 29].

- In coal mining-related operations throughout the United States, during the period 1990–1999, 197 equipment fires resulted in 76 injuries [30].
- As per Ref. [31], during the period of 1990–1994, around 27% of the commercial nuclear power plant outages in the United States were the result of human error.
- A Boeing study reported that approximately 19.2% of in-flight engine shutdowns are due to maintenance error [32].
- In 1979, in a DC-10 aircraft accident in Chicago, 272 persons lost their lives because of wrong procedures followed by maintenance personnel [33].
- In 1991, United Airlines Flight 585 (aircraft type: Boeing 737-291) crashed because of rudder device malfunction and caused 25 fatalities [34].
- In 2002, an Amtrak auto train derailed because of malfunctioning brakes and poor track maintenance near Crescent City, Florida, and caused four deaths and 142 injuries [35].
- As per Ref. [36], the Emergency Care Research Institute after examining a sample of 15,000 hospital products concluded that about 4%–6% of these products were dangerous enough for warranting immediate corrective measure [36].
- The Internet has grown from four hosts in 1969 to over 147 hosts and 38 sites in 2002, and in 2001, there were 52,000 Internet-related failures and incidents [37].

## 1.3 Terms and definitions

There are a large number of terms and definitions used in the area of reliability, maintainability, and safety. Some of these are presented in the following [4, 38–41]:

- **Reliability:** The probability that an item will perform its assigned mission satisfactorily for the stated period when used according to the specified conditions.
- **Maintainability:** The probability that a failed item will be restored to its satisfactory operational state.
- **Safety:** The conservation of human life and the prevention of damage to items as per mission requirements.
- **Maintenance:** All actions appropriate for retaining an item/equipment in, or restoring it to, a given condition.
- **Failure:** The inability of an item to function within the stated guidelines.
- **Mission time:** The element of uptime that is needed to perform a stated mission profile.

- **Useful life:** The length of time a product operates within a tolerable level of failure rate.
- **Continuous task:** A task that involves some kind of tracking activity (e.g., monitoring a changing condition/situation).
- **Downtime:** The time period during which the item/system is not a condition to conduct its specified mission.
- **Failure mode:** The abnormality of a system/item performance that causes the item/system to be considered as failed.
- **Preventive maintenance:** All actions conducted on a planned, periodic, and specific schedule for keeping an item/equipment in the stated operating condition through the process of checking and reconditioning. These actions are precautionary steps undertaken for lowering or forestalling the probability of failure or an acceptable level of degradations in later service rather than correcting them after their occurrence.
- **Safeguard:** A barrier guard, a device, or a procedure developed for protecting humans.
- **Safety management:** The accomplishment of safety through the effort of other personnel.
- **Overhaul:** A comprehensive inspection and restoration of a piece of equipment or an item to an acceptable level at a durability time or usage limit.
- **Serviceability:** The degree of ease or difficulty with which an item/equipment can be restored to its working condition.
- **Safety process:** A series of procedures followed for enabling all safety-related requirements of an item to be identified and satisfied.
- **Unsafe condition:** Any condition, under the right set of conditions, that will result in an accident.
- **Predictive maintenance:** The use of modern measurement and signal-processing approaches for accurately diagnosing equipment/item condition during operation.
- **Corrective maintenance:** The unscheduled repair/maintenance for returning items/equipment to a defined state and conducted because maintenance persons or users perceived failures/deficiencies.
- **Hazard:** The source of energy and the physiological and behavioral factors that, when uncontrolled properly, lead to harmful occurrences.
- **Redundancy:** The existence of more than one means for conducting a specified function.
- **Availability:** The probability that an item/system is available for application or use when needed.
- **Inspection:** The qualitative observation of an item's/system's condition/performance.
- **Accident:** This is an unplanned and undesired event.
- **Injury:** This is a wound or other specific/certain damage.

## 1.4 Useful sources for obtaining information on reliability, maintainability, and safety

There are many sources for obtaining information, directly or indirectly, concerned with engineering system reliability, maintainability, and safety. Some of the sources considered most useful are presented in the following sections under six distinct categories.

### 1.4.1 Organizations

- Reliability Society, Institute of Electrical and Electronics Engineers (IEEE), PO BOX 1331, Piscataway, New Jersey.
- Society for Maintenance and Reliability Professionals, 401 North Michigan Avenue, Chicago, Illinois.
- Society for Machinery Failure Prevention Technology, 4193 Sudley Road, Haymarket, Virginia.
- Japan Institute of Plant Maintenance, Shuwa Shiba-Koen 3-Chome Building, 3-1-38, Shiba-Koen, Minato-ku, Tokyo, Japan.
- American Society of Safety Engineers, 1800 East Oakton Street, Des Plaines, Illinois.
- System Safety Society, 1452 Culver Drive, Suite A-261, Irvine, California.
- National Safety Council, 444 North Michigan Avenue, Chicago, Illinois.
- British Safety Council, 62 Chancellors Road, London, United Kingdom.

### 1.4.2 Standards and reports

- MIL-HDBK-217, Reliability Prediction of Electronic Equipment, Department of Defense, Washington, D.C.
- MIL-STD-756, Reliability Modeling and Prediction, Department of Defense, Washington, D.C.
- MIL-STD-785, Reliability Program for Systems and Equipment, Development, and Production, Department of Defense, Washington, D.C.
- MIL-STD-1629, Procedures for Performing Failure Mode, Effects, and Criticality Analysis, Department of Defense, Washington, D.C.
- MIL-STD-2155, Failure Reporting, Analysis, and Corrective Action, Department of Defense, Washington, D.C.
- Guide to Reliability-Centered Maintenance, Report No. AMCP 705-2, Department of Army, Washington, D.C, 1985.
- NATO ARMP-1 ED2, NATO Requirements for Reliability and Maintainability, North Atlantic Treaty Organization (NATO), Brussels, Belgium.

- NATO ARMP-5 AMDO, Guidance on Reliability and Maintainability Training, North Atlantic Treaty Organization, Brussels, Belgium.
- MIL-HDBK-472, Maintainability Prediction, Department of Defense, Washington, D.C.
- AMCP 706-133, Maintainability Guide for Design, Department of Defense, Washington, D.C.
- Maintenance Engineering Techniques, Report No. AMCP 706-132, Department of the Army, Washington, D.C, 1975.
- DEF-STD-00-55-1, Requirements for Safety-related Software in Defense Equipment, Department of Defense, Washington, D.C.
- MIL-STD-882, Systems Safety Program for System and Associated Subsystem and Equipment-Requirements, Department of Defense, Washington, D.C.
- MIL-STD-58077, Safety Engineering of Aircraft System, Associated subsystem and equipment: General Requirements, Department of Defense, Washington, D.C.

### 1.4.3 *Data information sources*

- Government Industry Data Exchange Program (GIDEP), GIDEP Operations Center, U.S. Department of the Navy, Corona, California.
- IEC 706 PTS, Guide on Maintainability of Equipment, Part III: Sections Six and Seven, Verification and Collection, Analysis and Presentation of Data, First Edition, International Electro-Technical Commission (IEC), Geneva, Switzerland.
- Defense Technical Information Center, DTIC-FDAC, 8725 John J. Kingman Road, Suite 0944, Fort Belvoir, Virginia.
- RACEEMDI, Electronics Equipment Maintainability Data, Reliability Analysis Center, Rome Air Development Center, Griffiss Air Force Base, Rome, New York.
- National Technical Information Center, 5285 Port Royal Road, Springfield, Virginia.
- National Aeronautics and Space Administration (NASA) Parts Reliability Information Center, George C. Marshall Space Flight Center, Huntsville, Alabama.
- American National Standards Institute, 11 W 42nd Street, New York, New York 10036.
- Gertman, D.I., Blackman, H.S., Human Reliability and Safety Analysis Data Handbook, Wiley, New York, 1994.

### 1.4.4 *Journals and magazines*

- IEEE Transactions on Reliability
- International Journal of Reliability, Quality, and Safety Engineering

- Reliability Engineering and System Safety
- Microelectronics and Reliability
- Journal of Quality in Maintenance Engineering
- Reliability: The Magazine for Improved Plant Reliability
- Journal of Safety Research
- Accident Analysis and Prevention
- National Safety News
- RAMS ASIA (Reliability, Availability, Maintainability, and Safety [RAMS], Quarterly Journal)
- Safety Management Journal
- Industrial Maintenance and Plant Operation

### 1.4.5 Conference proceedings

- Proceedings of the Annual Reliability and Maintainability Symposium
- Proceedings of the European Conferences on Safety and Reliability
- Proceedings of the ISSAT International Conferences on Reliability and Quality in Design
- Proceedings of the International Conferences on Probabilistic Safety Assessment and Management
- Proceedings of the System Safety Conferences

### 1.4.6 Books

- Shooman, M.L., Probabilistic Reliability: An Engineering Approach, McGraw-Hill, New York, 1968.
- Cox, S.J., Reliability, Safety, and Risk Management: An Integrated Approach, Butterworth-Heinemann, New York, 1991.
- Dhillon, B.S., Design Reliability: Fundamentals and Applications, CRC Press, Boca Raton, Florida, 1999.
- Nakagawa, T., Maintenance Theory of Reliability, Springer Inc., London, 2005.
- Blanchard, B.S., Verma, D., Peterson, E.L., Maintainability: A Key to Effective Serviceability and Maintenance Management, John Wiley and Sons, New York, 1995.
- Dhillon, B.S., Engineering Maintainability: How to Design for Reliability and Easy Maintenance, Gulf Publishing, Houston, Texas, 1999.
- Goldman, A.S., Slattery, T.B., Maintainability, John Wiley and Sons, New York, 1964.
- Handley, W., Industrial Safety Handbook, McGraw-Hill, New York, 1969.

- Dhillon, B.S., Robot System Reliability and Safety: A Modern Approach, CRC Press, Boca Raton, Florida, 2015.
- Blanchard, B.S., Lowery, E.E., Maintainability Principles and Practices, McGraw-Hill, New York, 1969.
- Dhillon, B.S., Computer System Reliability: Safety and Usability, CRC Press, Boca Raton, Florida, 2013.
- August, J., Applied Reliability-Centered Maintenance, Penn Well, Tulsa, Oklahoma, 1999.
- Dhillon, B.S., Engineering Safety: Fundamentals, Techniques, and Applications, World Scientific Publishing, River Edge, New Jersey, 1996.
- Cunningham, C.E., Cox, W., Applied Maintainability Engineering, John Wiley and Sons, New York, 1972.
- Dhillon, B.S., Mining Equipment Reliability, Maintainability, and Safety, Springer Inc., London, 2008.
- Smith, D.J., Babb, A.H., Maintainability Engineering, Pitman, New York, 1973.
- Dhillon, B.S., Maintainability, Maintenance, and Reliability for Engineers, CRC Press, Boca Raton, Florida, 2006.
- Dhillon, B.S., Transportation Systems Reliability and Safety, CRC Press, Boca Raton, Florida, 2011.
- Stephans, R.A., Talso, W.W., Editors, System Safety Analysis Handbook, System Safety Society, Irvine, California, 1993.


## 1.5 Scope of the book

Nowadays, engineering systems are an important element of world economy, and each year, a vast sum of money is spent for developing, manufacturing, operating, and maintaining various types of engineering systems around the globe. Global competition and other factors are clearly forcing manufacturers to produce highly reliable, safe, and maintainable engineering systems/products. Over the years, a large number of journal and conference proceeding articles, technical reports, etc., on the reliability, the maintainability, and the safety of engineering systems have appeared in the literature. However, to the best of the author's knowledge, there is no book that covers the topics of reliability, maintainability, and safety within its framework. This is a significant impediment to information seekers on these three topics because they have to consult various sources.

Thus, the main objectives of this book are (a) to eliminate the need for professionals and others concerned with engineering system reliability, maintainability, and safety to consult various different and diverse sources in obtaining desired information and (b) to provide up-to-date information on the topic. The book will be useful to many individuals including reliability, maintainability, and safety professionals concerned with engineering systems, engineering system administrators, engineering undergraduate and graduate students, instructors and researchers in the area of engineering systems, and engineers at large.

## 1.6 Problems

1. Write an essay on engineering system reliability, maintainability, and safety.
2. List at least six important facts and figures concerning engineering system reliability, maintainability, and safety.
3. Define the following three terms:
   i. Reliability.
   ii. Maintainability.
   iii. Safety.
4. List three examples concerning engineering system reliability/maintainability/safety-related problems.
5. List at least five, directly or indirectly, engineering system reliability/maintainability/safety data information sources.
6. Define the following four terms:
   i. Safety management.
   ii. Safety process.
   iii. Preventive maintenance.
   iv. Continuous task.
7. What is the difference between the terms predictive maintenance and corrective maintenance?
8. List five important organizations for obtaining information related to engineering system reliability, maintainability, and safety.
9. List at least eight standards/reports concerned with reliability/maintainability/safety.
10. What is the difference between reliability and availability concerning engineering systems.

## References

1. Layman, W.J., Fundamental Considerations in Preparing a Master System Plan, Electrical World, Vol. 101, 1933, pp. 778–792.
2. Smith, S.A., Service Reliability Measured by Probabilities of Outage, Electrical World, Vol. 103, 1934, pp. 371–374.
3. Dhillon, B.S., Reliability and Quality Control: Bibliography on General and Specialized Areas, Beta Publishers, Gloucester, Ontario, Canada, 1992.
4. AMCP 706-133, Engineering Design Handbook: Maintainability Engineering Theory and Practice, Department of Defense, Washington, D.C., 1976.
5. Moss, M.A., Minimal Maintenance Expense, Marcel Dekker, New York, 1985.

6. Shooman, M.L., Probabilistic Reliability: An Engineering Approach, McGraw-Hill, New York, 1968.
7. Dhillon, B.S., Engineering Maintainability: How to Design for Reliability and Easy Maintenance, Gulf Publishing, Houston, Texas, 1999.
8. Dhillon, B.S., Maintainability, Maintenance, and Reliability for Engineers, CRC Press, Boca Raton, Florida, 2006.
9. Goetsch, D.L., Occupational Safety and Health, Prentice Hall, Englewood Cliffs, New Jersey, 1996.
10. Hammer, W., Price, D., Occupational Safety Management and Engineering, Prentice Hall, Upper Saddle River, New Jersey, 2001.
11. Dhillon, B.S., Engineering Safety: Fundamentals, Techniques, and Applications, World Scientific Publishing, River Edge, New Jersey, 2003.
12. Herrmann, D.S., Software Safety and Reliability, IEEE Computer Society Press, Los Alamitos, California, 1999.
13. Kletz, T., Reflections on Safety, Safety Systems, Vol. 6, No. 3, 1997, pp. 1–3.
14. Latino, C.J., Hidden Treasure: Eliminating Chronic Failures Can Cut Maintenance Costs up to 60%, Report, Reliability Center, Hopewell, Virginia, 1999.
15. Trager, T.A., Case Study Report on Loss of Safety System Function Events, Report No. AEOD/C 504, United States Nuclear Regulatory Commission (NRC), Washington, D.C., 1985.
16. National Institute of Standards and Technology (NIST), 100 Bureau Drive, Stop 1070, Gaithersburg, Maryland, 2002.
17. Backtrom, T., Dooes, M.A., A Comparative Study of Occupational Accidents in Industries with Advanced Manufacturing Technology, International Journal of Human Factors in Manufacturing, Vol. 5, 1995, pp. 267–292.
18. Clark, D.R., Lehto, M.D., Reliability, Maintenance, and Safety of Robots, in Handbook of Industrial Robotics, edited by S.Y. Nof, Wiley, New York, 1999, pp. 717–753.
19. Accident Facts, National Safety Council, Chicago, 1990–1993.
20. "How to Reduce Workplace Accidents", Report, European Agency for Safety and Health at Work, Brussels, Belgium, 2001.
21. Banta, H.D., The Regulation of Medical Devices, Preventive Medicine, Vol. 19, 1990, pp. 693–699.
22. Medical Devices, Hearing Before the Sub-Committee on Public Health and Environment, US Congress Interstate and Foreign Commerce, Serial No. 93-61, US Government Printing Office, Washington, D.C., 1973.
23. Retsch, T., Schmitter, G., Marty, A., Safety Principles for Industrial Robots, in Encyclopedia of Occupational Health and Safety, Vol. II, edited by J.M., Stellman, International Labor Organization, Geneva, Switzerland, 2011, pp. 58.56–58.58.
24. Krauz, D.C., Gramopadhys, A.K., Effect of Team Training on Aircraft Maintenance Technicians: Computer-Based Training versus Instructor-Based Training, International Journal of Industrial Ergonomics, Vol. 27, 2001, pp. 141–157.
25. Phillips, E.H., Focus on Accident Prevention Key to Future Airline Safety, Aviation Week and Space Technology, Vol. 141, No. 9, 1994, pp. 52–53.
26. Human Factors in Airline Maintenance: A Study of Incident Reports, Bureau of Air Safety Inspection, Department of Transport and Regional Development, Canberra, Australia, 1997.

27. Russell, P.D., Management Strategies for Accident Prevention, Air Asia, Vol. 6, 1994, pp. 31–41.
28. Pyy, P., An Analysis of Maintenance Failures at a Nuclear Power Plant, Reliability Engineering and System Safety, Vol. 72, 2001, pp. 293–302.
29. Pyy, P., Laakso, K., Reiman, L., A Study of Human Errors Related to NPP Maintenance Activities, Proceedings of the IEEE Sixth Annual Human Factors Meeting, 1997, pp. 12.23–12.28.
30. De Rosa, M., Equipment Fires Causes Injuries: NIOSH Study Reveals Trends for Equipment Fires at US Coal Mines, Coal Age, October 2004, pp. 28–31.
31. Varma, V., Maintenance Training Reduces Human Errors, Power Engineering, Vol. 100, 1996, pp. 44–47.
32. Marx, D.A., Learning from Our Mistakes: A Review of Maintenance Error Investigation and Analysis Systems (with Recommendations to FAA), Federal Aviation Administration (FAA), Washington, D.C., January 1998.
33. Christensen, J.M., Howard, J.M., Field Experience in Maintenance, in Human Detection and Diagnosis of System Failures, edited by W. Karwowski, et al., Elsevier, Amsterdam, Netherlands, 1988, pp. 391–396.
34. Aircraft Accident Report: United Airlines Flight 585, Report No. AAR92-06, National Transportation Safety Board (NTSB), Washington, D.C., 1992.
35. Derailment of Amtrak Auto Train P052-18 on the CSXT Railroad Near Crescent City, Florida, Report No. RAR-03/02, National Transportation Safety Board, Washington, D.C., 2003.
36. Dhillon, B.S., Reliability Technology in Healthcare Systems, Proceedings of the IASTED International Symposium on Computers, Advanced Technology in Medicine, and Health Care Bioengineering, 1990, pp. 84–87.
37. Hafner, K., Lyon, M., Where Wizards Stay Up Late: The Origin of the Internet, Simon and Schuster, New York, 1996.
38. MIL-STD-721, Definitions of Effectiveness Terms for Reliability, Maintainability, Human Factors, and Safety, Department of Defense, Washington, D.C.
39. Omdahl, T.P., Editor, Reliability, Availability, Maintainability (RAM) Dictionary, ASQC Quality Press, Milwaukee, Wisconsin, 1988.
40. Dictionary of Terms Used in the Safety Profession, American Society of Safety Engineers (ASEE), 3rd Edition, Des Plaines, Illinois, 1988.
41. McKenna, T., Oliverson, R., Glossary of Reliability and Maintenance Terms, Gulf Publishing, Houston, Texas, 1997.

# chapter two

# Reliability, maintainability, and safety mathematics

## 2.1 Introduction

As in the development of other areas of science and technology, mathematics has also played an important role in the development of reliability, maintainability, and safety fields as well. The history of mathematics may be traced back to the development of our currently used number symbols, sometimes in the published literature referred to as the 'Hindu-Arabic numeral system' [1]. Among the early evidences of these number symbols' use are notches found on stone columns erected by the Scythian Emperor of India named Asoka, in around 250 BC [1].

The earliest reference to the probability concept may be traced back to the gambler's manual written by Girolamo Cardano (1501–1576) [2]. However, Pierre Fermat (1601–1665) and Blaise Pascal (1623–1662) were the first two persons who solved independently and correctly the problem of dividing the winnings in a game of chance [1, 2]. Boolean algebra, which plays a key role in modern probability theory, is named after an English mathematician George Boole (1815–1864), who published in 1847 a pamphlet titled 'The Mathematical Analysis of Logic: Being an Essay towards a Calculus of Deductive Reasoning' [1–3].

Laplace transforms, often used in the reliability area for finding solutions to first-order differential equations, were developed by a French mathematician named Pierre-Simon Laplace (1749–1827). Additional information on the history of mathematics and probability is available in Refs. [1, 2].

This chapter presents basic mathematical concepts considered useful to understand subsequent chapters of this book.

## 2.2 Arithmetic mean and mean deviation

A given set of data related to reliability/maintainability/safety of engineering systems is useful only if it is analyzed properly. More clearly, there are certain characteristics of the data that are useful for describing the nature of a given dataset, thus making better related decisions.

Thus, this section presents two statistical measures considered useful for studying reliability, maintainability, and safety-related data for engineering systems [4, 6].

### 2.2.1 *Arithmetic mean*

This is defined by

$$m = \frac{\sum_{i=1}^{k} x_i}{k} \tag{2.1}$$

where

m is the mean value (i.e., arithmetic mean).
$x_i$ is the data value i, for i = 1, 2, 3, . . . , k.
k is the number of data values.

It is to be noted that the arithmetic mean is generally called 'mean'.

**Example 2.1**

Assume that the quality control department of an engineering system manufacturing company inspected six identical engineering systems and found 2, 4, 6, 7, 3, and 2 defects in each respective engineering system. Calculate the average number of defects (i.e., arithmetic mean) per engineering system.

By substituting the specified data values into Equation (2.1), we get

$$m = \frac{2+4+6+7+3+2}{6} = 4 \text{ defects per engineering system}$$

Thus, the average number of defects per engineering system is 4. In other words, the arithmetic mean of the given dataset is 4.

### 2.2.2 *Mean deviation*

This is a measure of dispersion whose value indicates the degree to which a given set of data tends to spread about a mean value and is expressed by

$$MD = \frac{\sum_{i=1}^{n} |x_i - m|}{n} \tag{2.2}$$

where

MD is the mean deviation.
m is the mean value of the given dataset.
n is the number of data values.
$x_i$ is the data value i, for i = 1, 2, 3, . . . , n.
$|x_i - m|$ is the absolute value of the deviation of $x_i$ from m.

**Example 2.2**

Calculate the mean deviation of the dataset given in Example 2.1.

In Example 2.1, the calculated mean value (i.e., arithmetic mean) of the given dataset is 4 defects per engineering system. Thus, by using this calculated value and the given data values in Equation (2.2), we get

$$MD = \frac{|2-4| + |4-4| + |6-4| + |7-4| + |3-4| + |2-4|}{6}$$

$$= \frac{[2+0+2+3+1+2]}{6}$$

$$= 1.66$$

Thus, the mean deviation of the dataset in Example 2.1 is 1.66.

## 2.3 *Boolean algebra laws*

Boolean algebra, named after its founder, George Boole (1815–1864), is used to a degree in the studies of reliability, maintainability, and safety related in engineering systems. Some of its laws that are considered useful for understanding subsequent chapters of this book are as follows [3, 5–8]:

$$A.B = B.A \tag{2.3}$$

where

A is an arbitrary set or event.
B is an arbitrary set or event.
Dot (.) denotes the intersection of sets.

It is to be noted that sometimes Equation (2.3) or other similar Equations are written without the dot (e.g., AB), but they still convey the same meaning.

$$A + B = B + A \tag{2.4}$$

where
+ is the union of sets or events.

$$A + A = A \tag{2.5}$$

$$AA = A \tag{2.6}$$

$$B(B + A) = B \tag{2.7}$$

$$A + AB = A \tag{2.8}$$

$$A(B + C) = AB + AC \tag{2.9}$$

where
C is an arbitrary set or event.

$$(A + B)(A + C) = A + BC \tag{2.10}$$

$$(A + B) + C = A + (B + C) \tag{2.11}$$

$$(AB)C = A(BC) \tag{2.12}$$

It is to be noted that in the published literature, Equations (2.3) and (2.4) are called commutative law, Equations (2.5) and (2.6) idempotent law, Equations (2.7) and (2.8) absorption law, Equations (2.9) and (2.10) distributive law, and Equations (2.11) and (2.12) associative law [9].

## 2.4 *Probability definition and properties*

The probability is defined as follows [10]:

$$P(X) = \lim_{n \to \infty} \left( \frac{N}{n} \right) \tag{2.13}$$

where
P(X) is the occurrence probability of event X.
N is the number of times event X occurs in the n repeated experiments.

Some of the basic probability properties are as follows [7, 10]:

- The probability of the occurrence of an event, say event X, is

$$0 \le P(X) \le 1 \tag{2.14}$$

- The probability of the occurrence and nonoccurrence of an event, say event X, is always:

$$P(X)+P(\bar{X})=1 \tag{2.15}$$

where
P(X) is the probability of the occurrence of event X.
$P(\bar{X})$ is the probability of the nonoccurrence of event $\bar{X}$.

- The probability of an intersection of n independent events is

$$P(X_1X_2X_3...X_n)=P(X_1)P(X_2)P(X_3)...P(X_n) \tag{2.16}$$

where
$P(X_i)$ is the occurrence probability of event $X_i$, for i = 1, 2, 3, . . . , n.

- The probability of the union of n independent events is

$$P(X_1+X_2+---+X_n)=1-\prod_{i=1}^{n}(1-P(X_i)) \tag{2.17}$$

- The probability of the union of n mutually exclusive events is

$$P(X_1+X_2+---+X_n)=\sum_{i=1}^{n}P(X_i) \tag{2.18}$$

**Example 2.3**

Assume that an engineering system is composed of two independent subsystems $X_1$ and $X_2$. The failure of either subsystem can result in engineering system failure. The probability of failure of subsystems $X_1$ and $X_2$ is 0.04 and 0.05, respectively.

Calculate the probability of failure of the engineering system.

By substituting the given data values into Equation (2.17), we get

$$\begin{aligned}P(X_1+X_2)&=1-\prod_{i=1}^{2}(1-P(X_i))\\&=P(X_1)+P(X_2)-P(X_1)P(X_2)\\&=0.04+0.05-(0.04)(0.05)\\&=0.088\end{aligned}$$

Thus, the probability of failure of the engineering system is 0.088.

## 2.5 Mathematical definitions

This section presents a number of definitions considered useful to perform various types of reliability, maintainability, and safety studies concerned with engineering systems.

### 2.5.1 Cumulative distribution function

For a continuous random variable, the cumulative distribution function is defined by [9, 10].

$$F(t) = \int_{-\infty}^{\infty} f(y)dy \tag{2.19}$$

where

y is a continuous random variable.
f(y) is the probability density function.
F(t) is the cumulative distribution function.

For $t = \infty$, Equation (2.19) becomes

$$F(\infty) = \int_{-\infty}^{\infty} f(y)dy \tag{2.20}$$
$$= 1$$

It means that the total area under the probability density curve is equal to unity.

Generally, in reliability, maintainability, and safety studies of engineering systems, Equation (2.19) is simply written as

$$F(t) = \int_{0}^{t} f(y)dy \tag{2.21}$$

**Example 2.4**

Assume that the probability (i.e., failure) density function of an engineering system is

$$f(t) = \alpha e^{-\alpha t}, \quad for\ t \geq 0, \alpha > 0 \tag{2.22}$$

where

f(t) is the probability density function (usually, in the area of reliability, it is called the failure density function).
t is a continuous random variable (i.e., time).
$\alpha$ is engineering system failure rate.

Obtain an expression for the engineering system cumulative distribution function by using Equation (2.22).

By substituting Equation (2.22) into Equation (2.21), we obtain

$$F(t) = \int_0^t \alpha e^{-\alpha t} dt \tag{2.23}$$
$$= 1 - e^{-\alpha t}$$

Thus, Equation (2.23) is the expression for the engineering system cumulative distribution function.

### 2.5.2 *Probability density function*

For a continuous random variable, the probability density function is expressed by [10]

$$f(t) = \frac{dF(t)}{dt} \tag{2.24}$$

where

f(t) is the density function.

F(t) is the cumulative distribution function.

**Example 2.5**

Prove with the aid of Equation (2.23) that Equation (2.22) is the probability density function.

By inserting Equation (2.23) into Equation (2.24), we obtain

$$f(t) = \frac{d\left(1 - e^{-\alpha t}\right)}{dt} \tag{2.25}$$
$$= \alpha e^{-\alpha t}$$

Equations (2.22) and (2.25) are identical.

### 2.5.3 *Expected value*

The expected value of a continuous random variable is expressed by [10]

$$E(t) = \int_{-\infty}^{\infty} t f(t) dt \tag{2.26}$$

where

E(t) is the expected value (i.e., mean value) of the continuous random variable t.

**Example 2.6**

Find the expected value (i.e., mean value) of the probability (failure) density function defined by Equation (2.22).

By inserting Equation (2.22) into Equation (2.26), we obtain

$$E(t) = \int_0^{\infty} t e^{-\alpha t} dt$$

$$= \left[ -t e^{-\alpha t} \right]_0^{\infty} - \left[ -\frac{e^{-\alpha t}}{\alpha} \right]_0^{\infty} \quad (2.27)$$

$$= \frac{1}{\alpha}$$

Thus, the expected value (i.e., mean value) of the probability (failure) density function defined by Equation (2.22) is given by Equation (2.27).

### 2.5.4 *Laplace transform definition and Laplace transforms of common functions*

The Laplace transform (named after a French Mathematician, Pierre-Simon Laplace [1749–1827]) of a function, say f(t), is expressed by [1, 11, 12]

$$f(s) = \int_0^{\infty} f(t) e^{-st} dt \quad (2.28)$$

where

s is the Laplace transform variable.
t is a variable.
f(s) is the Laplace transform of function f(t).

**Example 2.7**

Obtain the Laplace transform of the following function:

$$f(t) = e^{-\theta t} \quad (2.29)$$

where

$\theta$ is a constant.
t is a continuous random variable.

**Table 2.1** Laplace transforms of some functions

| No. | f(t) | f(s) |
|---|---|---|
| 1 | c (a constant) | $\frac{c}{s}$ |
| 2 | T | $\frac{1}{s^2}$ |
| 3 | $t^n$, *for* m = 0, 1, 2, 3, … | $\frac{m!}{s^{m+1}}$ |
| 4 | $e^{-\lambda t}$ | $\frac{1}{s+\lambda}$ |
| 5 | $te^{-\lambda t}$ | $\frac{1}{(s+\lambda)^2}$ |
| 6 | t f(t) | $-\frac{df(s)}{ds}$ |
| 7 | $\theta_1 f_1(t)+\theta_2 f_2(t)$ | $\theta_1 f_1(s)+\theta_2 f_2(s)$ |
| 8 | $\frac{df(t)}{dt}$ | s f(s) – f(0) |

By inserting Equation (2.29) into Equation (2.28), we get

$$f(s)=\int_0^\infty e^{-\theta t}e^{-st}dt$$

$$=\left[\frac{e^{-(s+\theta)t}}{(s+\theta)}\right]_0^\infty \tag{2.30}$$

$$=\frac{1}{s+\theta}$$

Thus, Equation (2.30) is the Laplace transform of Equation (2.29).

Laplace transforms of some commonly occurring functions in engineering system reliability, maintainability, and safety analysis studies are presented in Table 2.1 [11–13].

### 2.5.5 *Final value theorem Laplace transform*

If the following limits exist, then the final value theorem may be expressed as

$$\lim_{t\to\infty} f(t)=\lim_{s\to 0}\left[sf(s)\right] \tag{2.31}$$

**Example 2.8**

Prove, by using the following equation, that the left-hand side of Equation (2.31) is equal to its right side:

$$f(t)=\frac{\alpha_1}{(\alpha_1+\alpha_2)}+\frac{\alpha_2}{(\alpha_1+\alpha_2)}e^{-(\alpha_1+\alpha_2)t} \tag{2.32}$$

where

$\alpha_1$ and $\alpha_2$ are the constants.

By inserting Equation (2.32) into the left-hand side of Equation (2.31), we obtain

$$\lim_{t\to\infty}\left[\frac{\alpha_1}{\alpha_1+\alpha_2}+\frac{\alpha_2}{(\alpha_1+\alpha_2)}e^{-(\alpha_1+\alpha_2)t}\right]=\frac{\alpha_1}{(\alpha_1+\alpha_2)} \tag{2.33}$$

With the aid of Table 2.1, we get the following Laplace transforms of Equation (2.32):

$$f(s)=\frac{\alpha_1}{s(\alpha_1+\alpha_2)}+\frac{\alpha_1}{(\alpha_1+\alpha_2)}\cdot\frac{1}{(s+\alpha_1+\alpha_2)} \tag{2.34}$$

By substituting Equation (2.34) into the right hand of Equation (2.31), we get

$$\lim_{s\to 0} s\left[\frac{\alpha_1}{s(\alpha_1+\alpha_2)}+\frac{\alpha_2}{(\alpha_1+\alpha_2)}\frac{1}{(s+\alpha_1+\alpha_2)}\right]=\frac{\alpha_1}{(\alpha_1+\alpha_2)} \tag{2.35}$$

As the right-hand sides of Equations (2.33) and (2.35) are identical, it proves that the left-hand side of Equation (2.31) is equal to its right side.

## 2.6 *Probability distributions*

Although there are a large number of probability or statistical distributions in published literature, this section presents just four such distributions considered useful for performing various types of reliability, maintainability, and safety analysis studies concerned with engineering systems [14–16].

### 2.6.1 *Exponential distribution*

This is one of the simplest continuous random variable distributions frequently used in the industrial sector, particularly for performing reliability studies. Its probability density function is expressed by [9, 17]

$$f(t)=\alpha e^{-\alpha t}, \quad for\ t\geq 0, \alpha>0 \tag{2.36}$$

where
$f(t)$ is the probability density function.
$t$ is the time (i.e., a continuous random variable).
$\alpha$ is the distribution parameter.

By inserting Equation (2.36) into Equation (2.21), we get the equation for the cumulative distribution function:

$$F(t) = 1 - e^{-\alpha t} \tag{2.37}$$

With the aid of Equations (2.26) and (2.36), we obtain the following equation for the distribution expected value (i.e., mean value):

$$E(t) = \frac{1}{\alpha} \tag{2.38}$$

**Example 2.9**

Assume that the mean time to failure of engineering system is 1000 hours. Calculate the failure probability of the engineering system during a 500-hour mission by using Equations (2.37) and (2.38).

By inserting the specified data value into Equation (2.38), we obtain

$$\alpha = \frac{1}{1000} = 0.001 \text{ failures per hour}$$

By substituting the calculated and the given data values into Equation (2.37), we get

$$F(500) = e^{-(0.001)(500)}$$
$$= 0.6065$$

Thus, the failure probability of the engineering system during the 500-hour mission is 0.6065.

## 2.6.2 *Rayleigh distribution*

This continuous random variable probability distribution is named after its founder, John Rayleigh (1842–1919) [1], and its probability density function is defined by

$$f(t) = \left(\frac{1}{\theta^2}\right) t e^{-(t/\theta)^2}, \quad \text{for } \theta > 0, t \geq 0 \tag{2.39}$$

where
$\theta$ is the distribution parameter.

By substituting Equation (2.39) into Equation (2.21), we obtain the following equation for the cumulative distribution function:

$$F(t) = 1 - e^{-(t/\theta)^2} \tag{2.40}$$

By inserting Equation (2.39) into Equation (2.26), we get the following equation for the distribution expected value:

$$E(t) = \theta\Gamma\left(\frac{3}{2}\right) \tag{2.41}$$

where

$\Gamma(.)$ is the gamma function and is defined by

$$\Gamma(n) = \int_0^\infty t^{n-1} e^{-t} dt, \quad for\ n > 0 \tag{2.42}$$

### 2.6.3 *Weibull distribution*

This continuous random variable distribution was developed by Waloddi Weibull, a Swedish professor in mechanical engineering in the early 1950s and its probability density function is defined [18]

$$f(t) = \frac{bt^{b-1}}{\theta^b} e^{-(t/\theta)^b}, \quad for\ t \geq 0, b > 0, \theta > 0 \tag{2.43}$$

where

b and $\theta$ are the distribution shape and scale parameters, respectively.

By inserting Equation (2.43) into Equation (2.21), we obtain the following equation for the cumulative distribution function:

$$F(t) = 1 - e^{-(t/\theta)^b} \tag{2.44}$$

By substituting Equation (2.43) into Equation (2.26), we obtain the following equation for the distribution expected value:

$$E(t) = \theta\Gamma\left(1 + \frac{1}{b}\right) \tag{2.45}$$

It is to be noted that for b = 1 and b = 2, the exponential and Rayleigh distributions are the special cases of this distribution, respectively.

### 2.6.4 Bathtub hazard rate curve distribution

This is another continuous random variable distribution, and it can represent bathtub-shaped, increasing and decreasing hazard rates. This distribution was developed in 1981 [19], and in the published literature by other authors around the world, it is generally referred to as the Dhillon distribution/law/model [20–39].

The probability density function of the distribution is expressed by [19]

$$f(t) = b\theta(\theta t)^{b-1} e^{-\left\{e^{(\theta t)^b} - (\theta t)^b - 1\right\}} \quad \textit{for } t \geq 0, \theta > 0, b > 0 \tag{2.46}$$

where

b and $\theta$ are the distribution shape and scale parameters, respectively.

By substituting Equation (2.46) into Equation (2.21), we obtain the following equation for cumulative distribution function:

$$F(t) = 1 - e^{-\left\{e^{(\theta t)^b} - 1\right\}} \tag{2.47}$$

It is to be noted that for b = 0.5, this probability distribution gives the bathtub-shaped hazard rate curve, and for b = 1, it gives the extreme value probability distribution. In other words, the extreme value probability distribution is the special case of this probability distribution at b = 1.

## 2.7 Solving first-order differential equations with Laplace transforms

Usually, Laplace transforms are used to find solutions to first-order linear differential equations in reliability, maintainability, and safety analysis-related studies of engineering systems. The example presented below demonstrates the finding of solutions to a set of linear first-order differential equations, describing a engineering system in regard to reliability and safety, using Laplace transforms.

**Example 2.10**

Assume that an engineering system can be in any of the three states: operating normally, failed safely, or failed unsafely. The following three first-order linear differential equations describe the engineering system under consideration:

$$\frac{dP_0(t)}{dt} + (\lambda_s + \lambda_u) P_0(t) = 0 \tag{2.48}$$

$$\frac{dP_1(t)}{dt} - \lambda_s P_0(t) = 0 \tag{2.49}$$

$$\frac{dP_2(t)}{dt} - \lambda_u P_0(t) = 0 \tag{2.50}$$

where

$\lambda_s$ is the engineering system constant safe failure rate.

$\lambda_u$ is the engineering system constant unsafe failure rate.

$P_i(t)$ is the probability that the engineering system is in state i at time, for i = 0 (operating normally), i = 1 (failed safely), and i = 2 (failed unsafely).

At time t = 0, $P_0(0) = 1$, $P_1(0) = 0$, and $P_2(0) = 0$.

Solve differential Equations (2.48)–(2.50) by using Laplace transforms.

Using Table 2.1, differential Equations (2.48)–(2.50), and the given initial conditions, we obtain

$$sP_0(s) - 1 + (\lambda_s + \lambda_u) P_0(s) = 0 \tag{2.51}$$

$$sP_1(s) - \lambda_s P_0(s) = 0 \tag{2.52}$$

$$sP_2(s) - \lambda_u P_0(s) = 0 \tag{2.53}$$

By solving Equations (2.51)–(2.53), we get

$$P_0(s) = \frac{1}{(s + \lambda_s + \lambda_u)} \tag{2.54}$$

$$P_1(s) = \frac{\lambda_s}{s(s + \lambda_s + \lambda_u)} \tag{2.55}$$

$$P_2(s) = \frac{\lambda_u}{s(s + \lambda_s + \lambda_u)} \tag{2.56}$$

By taking the inverse Laplace transforms of Equations (2.54)–(2.56), we obtain

$$P_0(t) = e^{-(\lambda_s + \lambda_u)t} \tag{2.57}$$

$$P_1(t) = \frac{\lambda_s}{(\lambda_s + \lambda_u)}\left[1 - e^{-(\lambda_s + \lambda_u)t}\right] \tag{2.58}$$

$$P_2(t) = \frac{\lambda_u}{(\lambda_s + \lambda_u)}\left[1 - e^{-(\lambda_s + \lambda_u)t}\right] \tag{2.59}$$

Thus, Equations (2.57)–(2.59) are the solutions to differential Equations (2.48)–(2.50).

## 2.8 Problems

1. Assume that the quality control department of an engineering system manufacturing company inspected five identical engineering systems and found 4, 1, 3, 5, and 2 defects in each respective engineering system. Calculate the average number of defects (i.e., arithmetic mean) per engineering system.
2. Calculate the mean deviation of the dataset given in Question 1.
3. Define probability.
4. What is idempotent law?
5. What are the basic properties of probability?
6. Define the following three items:
   i. Probability density function.
   ii. Cumulative distribution function.
   iii. Expected value of a continuous random variable.
7. Define Laplace transform.
8. Write down the probability density function for the exponential distribution.
9. Prove Equations (2.57)–(2.59) by using Equations (2.54)–(2.56).
10. What are the special case distributions of the bathtub hazard rate curve and Weibull distributions?

## References

1. Eves, H., An Introduction to the History of Mathematics, Holt, Reinhart and Winston, New York, 1976.
2. Owen, D.B., On the History of Statistics and Probability, Marcel Dekker, New York, 1976.
3. Lipschutz, S., Set Theory, McGraw-Hill, New York, 1964.
4. Dhillon, B.S., Reliability, Quality, and Safety for Engineers, CRC Press, Boca Raton, Florida, 2004.
5. Spiegel, M.R., Statistics, McGraw-Hill, New York, 1961.
6. Spiegel, M.R., Probability and Statistics, McGraw-Hill, New York, 1975.
7. Lipschutz, S., Probability, McGraw-Hill, New York, 1965.
8. Fault Tree Handbook, Report No. NUREG-0492, U.S. Nuclear Regulatory Commission, Washington, D.C., 1981.
9. Dhillon, B.S., Computer System Reliability: Safety and Usability, CRC Press, Boca Raton, Florida, 2013.
10. Mann, N.R., Schafer, R.E., Singpurwalla, N.P., Methods for Statistical Analysis of Reliability and Life Data, John Wiley and Sons, New York, 1974.
11. Spiegel, M.R., Laplace Transforms, McGraw-Hill, New York, 1965.
12. Oberhettinger, F., Baddii, L., Tables of Laplace Transforms, Springer-Verlag, New York, 1973.
13. Nixon, F.E., Handbook of Laplace Transformation: Fundamentals, Applications, Tables, and Examples, Prentice Hall, Englewood Cliffs, New Jersey, 1960.

14. Shooman, M.L., Probabilistic Reliability: An Engineering Approach, McGraw-Hill, New York, 1968.
15. Patel, J.K., Kapadia, C.H., Owen, D.H., Handbook of Statistical Distributions, Marcel Dekker, New York, 1976.
16. Dhillon, B.S., Reliability Engineering in Systems Design and Operation, Van Nostrand and Reinhold, New York, 1983.
17. Davis, D.J., Analysis of Some Failure Data, Journal of the American Statistical Association, Vol. 47, 1952, pp. 113–150.
18. Weibull, W., A Statistical Distribution Function of Wide Applicability, Journal of Applied Mechanics, Vol. 18, 1951, pp. 293–297.
19. Dhillon, B.S., Life Distributions, IEEE Transactions on Reliability, Vol. 10, 1981, pp. 457–460.
20. Baker, R.D., Non-parametric Estimation of the Renewal Function, Computers Operations Research, Vol. 20, No. 2, 1993, pp. 167–178.
21. Cabana, A., Cabana, E.M., Goodness-of-fit to the Exponential Distribution, Focused on Weibull Alternatives, Communications in Statistics-Simulation and Computation, Vol. 34, 2005, pp. 711–723.
22. Grane, A., Fortiana, J., A Directional Test of Exponentiality Based on Maximum Correlations, Metrika, Vol. 73, 2011, pp. 711–723.
23. Henze, N., Meintnis, S.G., Recent and Classical Tests for Exponentiality: A Partial Review with Comparisons, Metrika, Vol. 61, 2005, pp. 29–45.
24. Jammalamadaka, S.R., Taufer, E., Testing Exponentiality by Comparing the Empirical Distribution Function of the Normalized Spacings with that of the Original Data, Journal of Nonparametric Statistics, Vol. 15, No. 6, 2003, pp. 719–729.
25. Hollander, M., Laird, G., Song, K.S., Non-parametric Interference for the Proportionality Function in the Random Censorship Model, Journal of Nonparametric Statistics, Vol. 15, No. 2, 2003, pp. 151–169.
26. Jammalamadaka, S.R., Taufer, E., Use of Mean Residual Life in Testing Departures from Exponentiality, Journal of Nonparametric Statistics, Vol. 18, No. 3, 2006, pp. 277–292.
27. Kunitz, H., Pamme, H., The Mixed Gamma Ageing Model in Life Data Analysis, Statistical Papers, Vol. 34, 1993, pp. 303–318.
28. Kunitz., H., A New Class of Bathtub-Shaped Hazard Rates and Its Application in Comparison of Two Test-Statistics, IEEE Transactions on Reliability, Vol. 38, No. 3, 1989, pp. 351–354.
29. Meintanis, S.G., A Class of Tests for Exponentiality Based on a Continuum of Moment Conditions, Kybernetika, Vol. 45, No. 6, 2009, pp. 946–959.
30. Morris, K., Szynal, D., Goodness-of-Fit Tests Based on Characterizations Involving Moments of Order Statistics, International Journal of Pure and Applied Mathematics, Vol. 38, No. 1, 2007, pp. 83–121.
31. Na, M.H., Spline Hazard Rate Estimation Using Censored Data, Journal of KSIAM, Vol. 3, No. 2, 1999, pp. 99–106.
32. Morris, K., Szynal, D., Some U-Statistics in Goodness-of-Fit Tests Derived from Characterizations via Record Values, International Journal of Pure and Applied Mathematics, Vol. 4, No. 4, 2008, pp. 339–414.
33. Nam, K.H., Park, D.H., Failure Rate for Dhillon Model, Proceedings of the Spring Conference of the Korean Statistical Society, 1997, pp. 114–118.

34. Nimoto, N., Zitikis, R., The Atkinson Index, the Moran Statistic, and Testing Exponentiality, Journal of the Japan Statistical Society, Vol. 38, No. 2, 2008, pp. 187–205.
35. Nam, K.H., Chang, S.J., Approximation of the Renewal Function for Hjorth Model and Dhillon Model, Journal of the Korean Society for Quality Management, Vol. 34, No. 1, 2006, pp. 34–39.
36. Noughabi, H.A., Arghami, N.R., Testing Exponentiality Based on Characterizations of the Exponential Distribution, Journal of Statistical Computation and Simulation, Vol. 1, First, 2011, pp. 1–11.
37. Szynal, D., Goodness-of-Fit Tests Derived from Characterizations of Continuous Distributions, Stability in Probability, Banach Center Publications, Vol. 90, Institute of Mathematics, Polish Academy of Sciences, Warszawa, Poland, 2010, pp. 203–223.
38. Szynal, D., Wolynski, W., Goodness-of-Fit Tests for Exponentiality and Rayleigh Distribution, International Journal of Pure and Applied Mathematics, Vol. 78, No. 5, 2013, pp. 751–772.
39. Nam, K.H., Park, D.H., A Study on Trend Changes for Certain Parametric Families, Journal of the Korean Society for Quality Management, Vol. 23, No.3, 1995, pp. 93–101.

# chapter three

# Reliability, maintainability, and safety basics

## 3.1 Introduction

Nowadays, the reliability of engineering systems has become a quite challenging issue during the design process due to the increasing dependence of our daily lives and schedules on the proper functioning of these systems. Some examples of these systems are computers, nuclear power-generating reactors, automobiles, space satellites, and aircraft.

The alarmingly high operating and support costs of engineering systems/equipment, in part due to failures and subsequent repairs, are the prime reasons for emphasizing maintainability of engineering systems. Some examples of these costs are the expense of maintenance personnel and their training, repair parts, test and support equipment, maintenance instructions and data, training equipment, and maintenance facilities.

Nowadays, engineering systems have become highly complex and sophisticated, and the safety of these systems has become a challenging issue. Needless to say, over the years, various types of methods and approaches have been developed for improving safety, reliability, and maintainability of engineering systems/equipment in the field. This chapter presents various reliability, maintainability, and safety basics considered useful to understand the subsequent chapters of this book.

## 3.2 Bathtub hazard rate curve

This curve is usually used to describe the failure rate of engineering systems/equipment and is shown in Fig. 3.1. The curve is called the bathtub hazard rate curve because it resembles the shape of a bathtub.

As shown in Fig. 3.1, the curve is divided into three sections. These sections are called burn-in period, useful-life period, and wear-out period. During the burn-in period, the system/equipment/item hazard rate decreases with time t. Some of the reasons for the occurrence of failures during this period are substandard materials and workmanship,

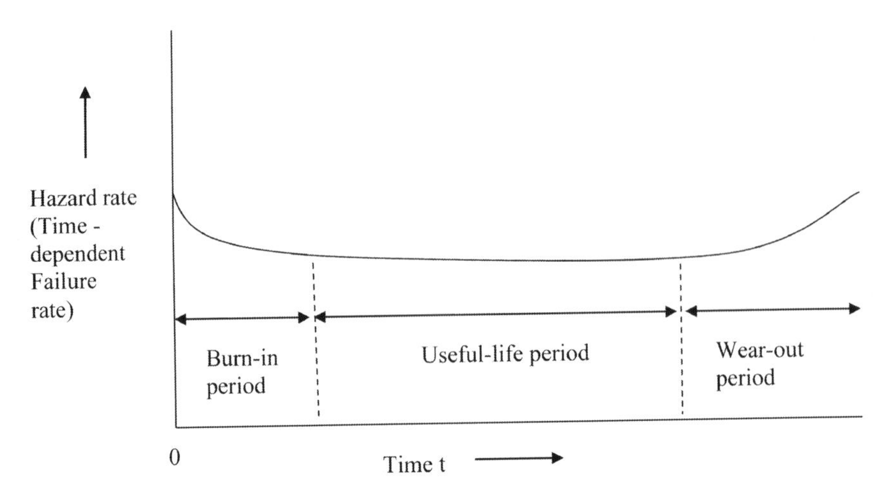


***Figure 3.1*** Bathtub hazard rate curve.

poor manufacturing methods and processes, inadequate debugging, poor quality control, and human error [1, 2]. Other terms used in the published literature from time to time for this region (period) are 'debugging region', 'infant-mortality region', and 'break-in region'.

During the useful life period, the hazard rate remains constant, and some of the causes for the occurrence of failures in this period are as follows [1, 2]:

- Higher random stress than expected.
- Undetectable defects.
- Natural failures.
- Abuse.
- Low safety factors.
- Human errors.

Finally, during the wear-out period, the hazard rate increases with time due to various reasons including poor maintenance practices, wear from aging, incorrect overhaul practices, wear due to friction, corrosion, and creep, and short designed-in life of the system/equipment/item under consideration [1, 2].

Mathematically, the following equation can be used to represent the bathtub hazard rate curve shown in Fig. 3.1 [3]:

$$\lambda(t) = \beta\alpha(\beta t)^{\alpha-1} e^{(\beta t)^{\alpha}} \tag{3.1}$$

where

$\lambda(t)$ is hazard rate (time-dependent failure rate).
t is time.
$\alpha$ is the shape parameter.
$\beta$ is the scale parameter.

Equation (3.1) at $\alpha = 0.5$ gives the shape of the bathtub hazard rate curve shown in Fig. 3.1.

## 3.3 General reliability formulas

A number of general formulas are often used for performing various types of reliability analysis. Four of these formulas that are based on the reliability function are as follows:

### 3.3.1 Failure (or probability) density function

The failure (or probability) density function is defined by [1]

$$f(t) = -\frac{dR(t)}{dt} \tag{3.2}$$

where

f(t) is the system/item failure (or probability) density function.
R(t) is the system/item reliability at time t.

**Example 3.1**

Assume that the reliability of an engineering system is expressed by

$$R_{es}(t) = e^{-\lambda_{es}t} \tag{3.3}$$

where

$R_{es}(t)$ is the engineering system reliability at time t.
$\lambda_{es}$ is the engineering system constant failure rate.

Obtain an expression for the failure (probability) density function of the engineering system by using Equation (3.2).

By inserting Equation (3.3) into Equation (3.2), we obtain

$$f(t) = -\frac{de^{-\lambda_{es}t}}{dt} = \lambda_{es}e^{-\lambda_{es}t} \tag{3.4}$$

Thus, Equation (3.4) is the expression for the failure (probability) density function of the engineering system.

### 3.3.2 *Hazard rate (or time-dependent failure rate) function*

This is defined by

$$\lambda(t) = \frac{f(t)}{R(t)} \tag{3.5}$$

where
$\lambda(t)$ is the system/item hazard rate (or time-dependent failure rate).

By inserting Equation (3.2) into Equation (3.5), we get

$$\lambda(t) = -\frac{1}{R(t)} \cdot \frac{dR(t)}{dt} \tag{3.6}$$

**Example 3.2**

Obtain an expression for the hazard rate of the engineering system by using Equations (3.3) and (3.6).

By substituting Equation (3.3) into Equation (3.6), we get

$$\begin{aligned} \lambda_{es}(t) &= -\frac{1}{e^{-\lambda_{es}t}} \cdot \frac{de^{-\lambda_{es}t}}{dt} \\ &= \lambda_{es} \end{aligned} \tag{3.7}$$

Thus, the hazard rate of the engineering system is given by Equation (3.7). It is to be noted that the right-hand side of this equation is not a function of time t. In other words, it is constant. Usually, it is called the constant failure rate of a system/item because it does not depend on time t.

### 3.3.3 *General reliability function*

This can be obtained with the aid of Equation (3.6). Thus, with the aid of Equation (3.6), we obtain

$$-\lambda(t)dt = \frac{1}{R(t)} \cdot dR(t) \tag{3.8}$$

By integrating the both sides of Equation (3.8) over the time interval [0, t], we obtain

$$-\int_0^t \lambda(t)dt = \int_1^{R(t)} \frac{1}{R(t)} \cdot dR(t) \tag{3.9}$$

Since, at time t = 0, R(t) = 1.

Evaluating the right-hand side of Equation (3.9) and then rearranging it yields

$$\ln R(t) = -\int_0^t \lambda(t)\,dt \tag{3.10}$$

Thus, from Equation (3.10), we obtain

$$R(t) = e^{-\int_0^t \lambda(t)\,dt} \tag{3.11}$$

Thus, Equation (3.11) is the general reliability function. This equation can be used for obtaining the reliability function of a system/item when its times to failure follow any time-continuous probability distribution (e.g., exponential, Weibull, and Rayleigh).

**Example 3.3**

Assume that the hazard rate of an engineering system is expressed by Equation (3.1). Obtain an expression for the reliability function of the engineering system by using Equation (3.1).

By substituting Equation (3.1) into Equation (3.11), we obtain

$$\begin{aligned} R(t) &= e^{-\int_0^t \left\{\beta\alpha(\beta t)^{\alpha-1} e^{(\beta t)^{\alpha}}\right\} dt} \\ &= e^{-\left\{e^{(\beta t)^{\alpha}} - 1\right\}} \end{aligned} \tag{3.12}$$

Thus, Equation (3.12) is the expression for the reliability function of the engineering system.

## 3.3.4 *Mean time to failure*

The mean time to failure of a system/item can be obtained by using any of the three formulas presented below [4, 5]:

$$MTTF = \int_0^\infty R(t)\,dt \tag{3.13}$$

or

$$MTTF = \lim_{s \to 0} R(s) \tag{3.14}$$

or

$$MTTF = E(t) = \int_0^\infty tf(t)\,dt \tag{3.15}$$

where

MTTF is the mean time to failure of a system/item.
s is the Laplace transform variable.
R(s) is the Laplace transform of the reliability function R(t).
E(t) is the expected value.

**Example 3.4**

With the aid of Equation (3.3), prove that Equations (3.13) and (3.14) yield the same result for the engineering system mean time to failure.

By substituting Equation (3.3) into Equation (3.13), we obtain

$$\begin{aligned} MTTF_{es} &= \int_0^\infty e^{-\lambda_{es} dt} \\ &= \frac{1}{\lambda_{es}} \end{aligned} \tag{3.16}$$

where

$MTTF_{es}$ is the engineering system mean time to failure.

By taking the Laplace transform of Equation (3.3), we get

$$\begin{aligned} R_{es}(s) &= \int_0^\infty e^{-st} e^{-\lambda_{es} t}\,dt \\ &= \frac{1}{(s+\lambda_{es})} \end{aligned} \tag{3.17}$$

where

$R_{es}(s)$ is the Laplace transform of the engineering system reliability function $R_{es}(t)$.

By inserting Equation (3.17) into Equation (3.14), we obtain

$$\begin{aligned} MTTF_{es} &= \lim_{s \to 0} \frac{1}{(s+\lambda_{es})} \\ &= \frac{1}{\lambda_{es}} \end{aligned} \tag{3.18}$$

As Equations (3.16) and (3.18) are identical, it proves that Equations (3.13) and (3.14) yield the same result for the engineering system mean time to failure.

## 3.4 Reliability networks

In analyzing reliability, the analyst can encounter parts of an engineering system forming a variety of different networks. Thus, this section is concerned with the reliability evaluation of such commonly occurring networks.

### 3.4.1 Series network

This is the simplest reliability network, and its block diagram is shown in Fig. 3.2. The diagram represents a k-unit system, and each block in the diagram denotes a unit. If any one of the k units fails, the series network/configuration/system fails. In other words, for the successful operation of the series system/network, all the k system/network units must function normally.

The series network/system, shown in Fig. 3.2, reliability is expressed by

$$R_{ss} = P(E_1 E_2 E_3 .... E_k) \tag{3.19}$$

where

$R_{ss}$ is the series system reliability.
$E_i$ is the successful operation (i.e., success event) of unit i; for i = 1,2,3, . . . ,k.
$P(E_1 E_2 E_3 \ldots E_k)$ is the occurrence probability of events $E_1 E_2 E_3 \ldots . E_k$.

For independently failing units, Equation (3.19) becomes

$$R_{ss} = P(E_1) P(E_2) P(E_3) ... P(E_k) \tag{3.20}$$

where

$P(E_i)$ is the probability of occurrence of event $E_i$, for i = 1,2,3, . . . ,k.

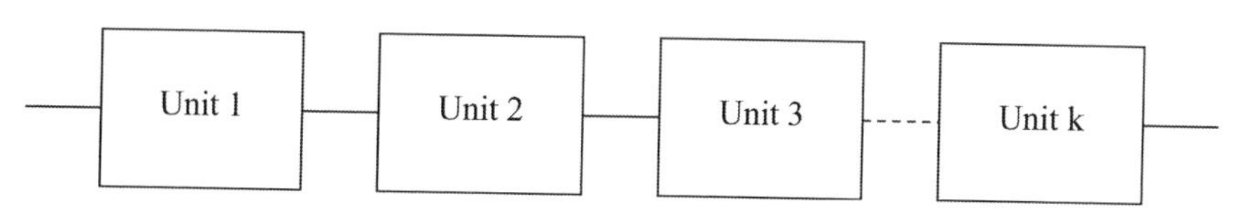


**Figure 3.2** Block diagram of a series network with k units.

If we let $R_i = P(E_i)$ for i = 1,2,3, . . . ,k, Equation (3.20) becomes

$$R_{ss} = R_1 R_2 R_3 ... R_k = \prod_{i=1}^{k} R_i \tag{3.21}$$

where
$R_i$ is the unit i reliability, for i = 1,2,3, . . . ,k.

For constant failure rate, $\lambda_i$, of unit i from Equation (3.11), we obtain

$$R_i(t) = e^{-\int_0^t \lambda_i dt} = e^{-\lambda_i t} \tag{3.22}$$

where
$R_i(t)$ is the reliability of unit i at time t.

By inserting Equation (3.22) into Equation (3.21), we get

$$R_{ss}(t) = e^{-\sum_{i=1}^{k} \lambda_i t} \tag{3.23}$$

where
$R_{ss}(t)$ is the series system reliability at time t.

By inserting Equation (3.23) into Equation (3.13), we obtain the following expression for the mean time to failure of the series system:

$$MTTF_{ss} = \int_0^\infty e^{-\sum_{i=1}^{k} \lambda_i t} dt = \frac{1}{\sum_{i=1}^{k} \lambda_i} \tag{3.24}$$

where
$MTTF_{ss}$ is the series system mean time to failure.

By substituting Equation (3.23) into Equation (3.6), we obtain the following expression for the series system hazard rate:

$$\lambda_{ss}(t) = -\frac{1}{e^{-\sum_{i=1}^{k}\lambda_i t}}\left[-\sum_{i=1}^{k}\lambda_i\right]e^{-\sum_{i=1}^{k}\lambda_i t}$$
$$= \sum_{i=1}^{k}\lambda_i \qquad (3.25)$$

where

$\lambda_{ss}(t)$ is the series system failure rate (hazard rate).

It is to be noted that right-hand side of Equation (3.25) is independent of time t. Thus, the left-hand side of Equation (3.25) is simply $\lambda_{ss}$, the failure rate of the series system. It means that whenever we add up failure rates of independent units/items, we automatically assume that these units/items form a series network, a worst-case design scenario with respect to reliability.

**Example 3.5**

Assume that an engineering system is composed of three independent and identical subsystems, and the constant failure rate of each subsystem is 0.0005 failures per hour. All three subsystems must operate normally for the engineering system to operate successfully.

Calculate the engineering system reliability for a 100-hour mission period, mean time to failure, and failure rate.

By substituting the specified data values into Equation (3.23), we get

$$R_{ss}(100) = e^{-(0.0005+0.0005+0.0005)(100)}$$
$$= 0.8607$$

Inserting the specified data values into Equation (3.24) yields

$$MTTF_{ss} = \frac{1}{(0.0005+0.0005+0.0005)}$$
$$= 666.67 \text{ hours}$$

By inserting the specified data values into Equation (3.25), we obtain

$$\lambda_{ss} = (0.0005+0.0005+0.0005) = 0.0015 \text{ failures per hour}$$

Thus, the engineering system reliability, mean time to failure, and failure rate are 0.8607, 666.67 hours, and 0.0015 failures per hour, respectively.

### 3.4.2 Parallel network

This network represents a system with k units/items operating simultaneously. For the system successful operation, at least one of the k units must operate normally. The block diagram of a k-unit parallel system is shown in Fig. 3.3, and each block in the diagram represents a unit.

The failure probability of the parallel system/network shown in Fig. 3.3 is expressed by

$$F_{ps} = P(\bar{x}_1\bar{x}_2\bar{x}_3...\bar{x}_k) \tag{3.26}$$

where

$F_{ps}$ is failure probability of the parallel system.
$\bar{x}_i$ is the failure (i.e., failure event) of unit i, for i = 1,2,3, . . . ,k.
$P(\bar{x}_1\bar{x}_2\bar{x}_3 \ldots \bar{x}_k)$ is the occurrence probability of events $\bar{x}_1, \bar{x}_2, \bar{x}_3, \ldots,$ and $\bar{x}_k$.

For independently failing parallel units, Equation (3.26) becomes

$$F_{ps} = P(\bar{x}_1)P(\bar{x}_2)P(\bar{x}_3)....P(\bar{x}_k) \tag{3.27}$$

where

$P(\bar{x}_i)$ is the occurrence probability of failure event $\bar{x}_i$, for i = 1,2,3, . . . ,k.

If we let $F_i = P(\bar{x}_i)$, for i = 1,2,3, . . . ,k, then Equation (3.27) becomes

$$F_{ps} = \prod_{i=1}^{k} F_i \tag{3.28}$$

where

$F_i$ is the failure probability of unit i, for i = 1,2,3 . . . ,k.

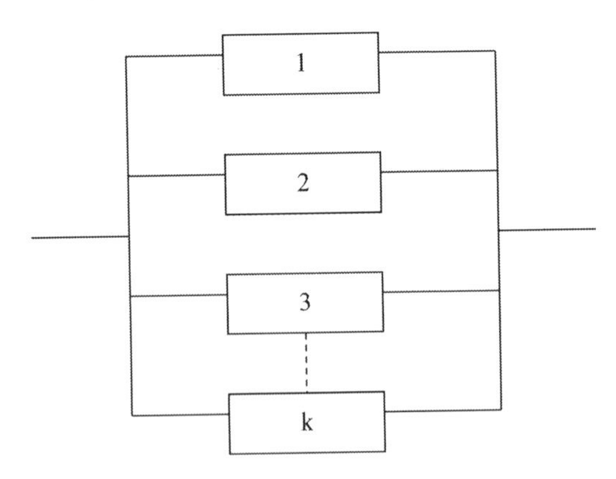


***Figure 3.3*** Block diagram of a parallel network/system with k units.

By subtracting Equation (3.28) from unity, we obtain

$$R_{ps} = 1 - \prod_{i=1}^{k} F_i \tag{3.29}$$

where

$R_{ps}$ is the parallel system/network reliability.

For constant failure rate $\lambda_i$ of unit i, subtracting Equation (3.22) from unity and then substituting it into Equation (3.29) yields

$$R_{ps}(t) = 1 - \prod_{i=1}^{k} \left(1 - e^{-\lambda_i t}\right) \tag{3.30}$$

where

$R_{ps}(t)$ is the parallel system/network reliability at time t.

For identical units, Equation (3.30) becomes

$$R_{ps}(t) = 1 - \left(1 - e^{-\lambda t}\right)^k \tag{3.31}$$

where

$\lambda$ is the unit constant failure rate.

By inserting Equation (3.31) into Equation (3.13), we obtain the following equation for the parallel system/network mean time to failure:

$$\begin{aligned} MTTF_{ps} &= \int_0^{\infty} \left[1 - \left(1 - e^{\lambda t}\right)^k\right] dt \\ &= \frac{1}{\lambda} \sum_{i=1}^{k} \frac{1}{i} \end{aligned} \tag{3.32}$$

where

$MTTF_{ps}$ is the identical units parallel system/network mean time to failure.

**Example 3.6**

Assume that an engineering system is composed of two independent, identical, and active units. At least one of the units must operate normally for the engineering system to function successfully. The failure rate of a unit is 0.0004 failures per hour.

Calculate the engineering system reliability for a 200-hour mission and mean time to failure.

By substituting the specified data values into Equation (3.31), we get

$$R_{ps}(200) = 1 - \left[1 - e^{-(0.0004)(200)}\right]^2$$
$$= 0.9940$$

Inserting the given data values into Equation (3.32) yields

$$MTTF_{ps} = \frac{1}{(0.0004)}\left[1 + \frac{1}{2}\right]$$
$$= 3750 \text{ hours}$$

Thus, engineering system reliability and mean time to failure are 0.9940 and 3750 hours, respectively.

### 3.4.3 *k-out-of-m network*

In this case, the network/system is composed of m active units, at least k units out of m active units must function normally for the successful system operation. The block diagram of a k-out-of-m unit network/system is shown in Fig. 3.4, and each block in the diagram represents a unit. It is to be noted that the series and parallel networks are special cases of this network for k = m and k = 1, respectively.

By using the binomial distribution, for identical and independent units, we write the following expression for reliability of k-out-of-m unit network shown in Fig. 3.4:

$$R_{k/m} = \sum_{i=k}^{m} \binom{m}{i} R^i (1 - R)^{m-i} \tag{3.33}$$

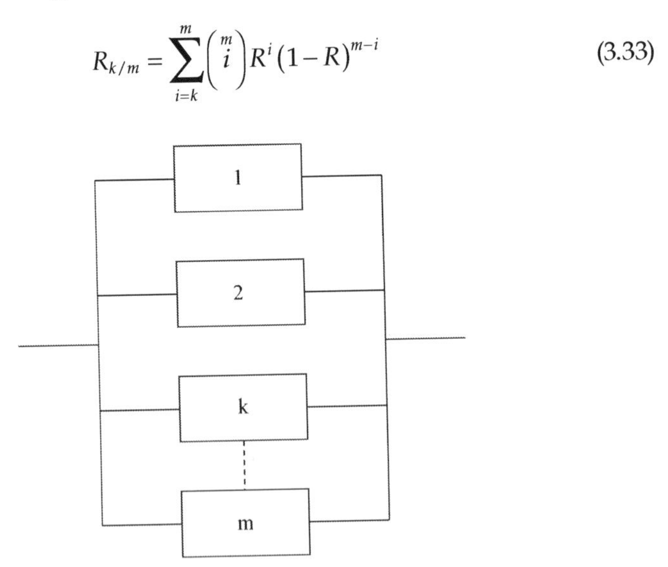


***Figure 3.4*** Block diagram of a k-out-of-m unit network/system.

where

$$\binom{m}{i} = \frac{m!}{(m-i)!i!} \tag{3.34}$$

$R_{k/m}$ is the k-out-of-m network/system reliability.
R is the unit reliability.

For constant failure rates of the identical units, by using Equations (3.11) and (3.33), we obtain

$$R_{k/m}(t) = \sum_{i=k}^{m} \binom{m}{i} e^{-i\lambda t} \left(1 - e^{-\lambda t}\right)^{m-i} \tag{3.35}$$

where
$R_{k/m}(t)$ is the k-out-of-m network/system reliability at time t.
$\lambda$ is the unit constant failure rate.

By substituting Equation (3.35) into Equation (3.13), we get

$$\begin{aligned} MTTF_{k/m} &= \int_0^{\infty} \left[ \sum_{i=k}^{m} \binom{m}{i} e^{-i\lambda t} (1 - e^{-\lambda t})^{m-i} \right] dt \\ &= \frac{1}{\lambda} \sum_{i=k}^{m} \frac{1}{i} \end{aligned} \tag{3.36}$$

where
$MTTF_{k/m}$ is the k-out-of-m network/system mean time to failure.

**Example 3.7**

Assume that an engineering system has three active, independent, and identical units in parallel. At least two units must operate normally for the successful operation of the engineering system. Calculate the engineering system mean time to failure if the unit constant failure rate is 0.0005 failures per hour.

By substituting the specified data values into Equation (3.36), we obtain

$$\begin{aligned} MTTF_{2/3} &= \frac{1}{(0.0005)} \left[ \frac{1}{2} + \frac{1}{3} \right] \\ &= 1666.66 \text{ hours} \end{aligned}$$

Thus, the engineering system mean time to failure is 1666.66 hours.

### 3.4.4 Standby system

This is another reliability network/system in which only one unit operates and k units are kept in their standby mode. The system is composed of (k + 1) units, and as soon as the operating unit fails, the switching mechanism detects the failure and turns on one of the standby units. The system fails when all its standby units fail.

The block diagram of a standby system with one operating and k standby units is shown in Fig. 3.5. Each block in the diagram represents a unit. By using Fig. 3.5, for identical and independent units, perfect switching mechanism and standby units, and time-dependent unit failure rate, we get the following expression for the standby system reliability [1, 6]:

$$R_{ss}(t) = \frac{\sum_{i=0}^{k}\left[\left[\int_0^t \lambda(t)dt\right]^i e^{-\int_0^t \lambda(t)dt}\right]}{i!} \tag{3.37}$$

where

$R_{ss}(t)$ is the standby system reliability at time t.
$\lambda(t)$ is the unit time-dependent failure rate or hazard rate.
k is the number of standby units.

For constant unit failure rate (i.e., $\lambda(t) = \lambda$), Equation (3.37) becomes

$$R_{ss}(t) = \frac{\sum_{i=0}^{k}(\lambda t)^i e^{-\lambda t}}{i!} \tag{3.38}$$

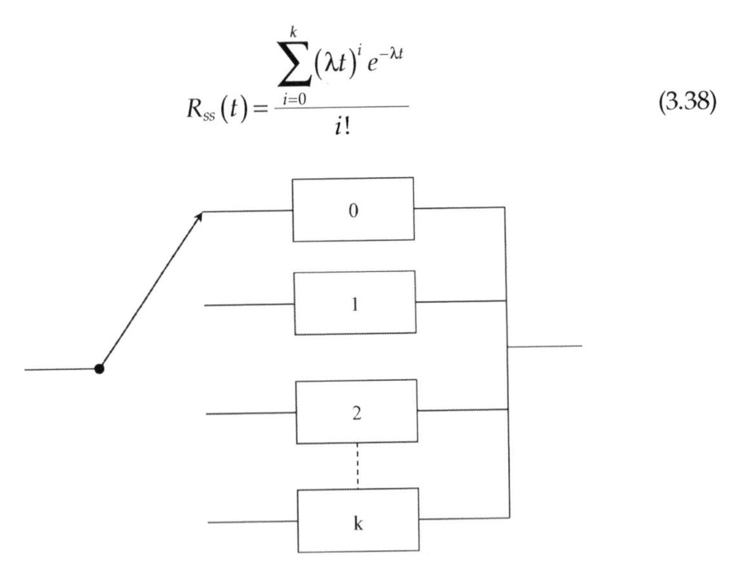


***Figure 3.5*** Block diagram of a standby system with one operating and k standby units.

where

$\lambda$ is the unit constant failure rate.

By inserting Equation (3.38) in Equation (3.13), we obtain

$$MTTF_{ss} = \int_0^\infty \left[ \frac{\sum_{i=0}^{k} (\lambda t)^i e^{-\lambda t}}{i!} \right] dt = \frac{(k+1)}{\lambda} \tag{3.39}$$

where

$MTTF_{ss}$ is the standby system mean time to failure.

**Example 3.8**

Assume that an engineering system is composed of a standby system having three independent and identical units: one operating and other two on standby. The unit constant failure rate is 0.0008 failures per hour.

Calculate the standby system mean time to failure if the switching mechanism is perfect and the standby units remain as good as new in their standby modes.

By inserting the given data values into Equation (3.39), we get

$$MTTF_{ss} = \frac{(2+1)}{(0.0008)} = 3750 \text{ hours}$$

Thus, the standby system mean time to failure is 3750 hours.

### 3.4.5 *Bridge network*

Sometimes units/parts in engineering systems may form a bridge network, as shown in Fig. 3.6. Each block in Fig. 3.6 diagram represents a unit/part, and all units/parts are labeled with numerals.

For independent units/parts, the bridge network shown in Fig. 3.6, reliability is expressed by [7]

$$\begin{aligned} R_{bn} = {} & 2R_1R_2R_3R_4R_5 + R_1R_3R_5 + R_2R_3R_4 + R_2R_5 + R_1R_4 \\ & - R_1R_2R_3R_4 - R_1R_2R_3R_5 - R_2R_3R_4R_5 - R_1R_2R_4R_5 \\ & - R_3R_4R_5R_1 \end{aligned} \tag{3.40}$$

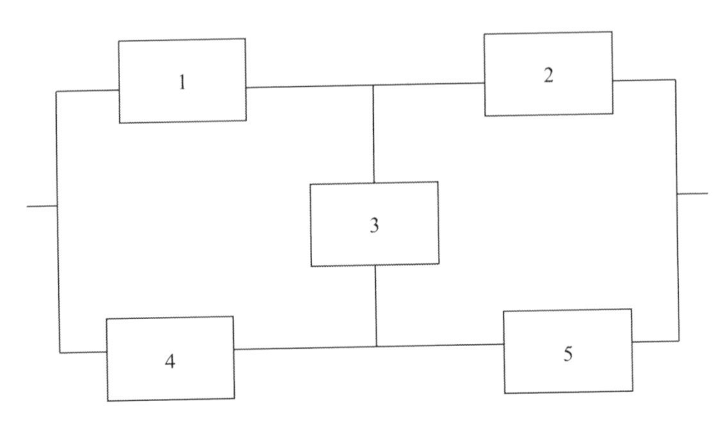


*Figure 3.6* Five nonidentical units bridge network.

where

$R_{bn}$ is the bridge network reliability.
$R_i$ is the unit i reliability, for i = 1,2,3,4,5.

For identical units, Equation (3.40) becomes

$$R_{bn} = R^5 - 5R^4 + 2R^3 + 2R^2 \quad (3.41)$$

where

R is the unit reliability.

For constant failure rates of all five units, and using Equations (3.11) and (3.41), we get

$$R_{bn}(t) = 2e^{-5\lambda t} - 5e^{-4\lambda t} + 2e^{-3\lambda t} + 2e^{-2\lambda t} \quad (3.42)$$

where

$R_{bn}(t)$ is bridge network reliability at time t.
$\lambda$ is the unit constant failure rate.

By substituting Equation (3.42) into Equation (3.13), we get

$$MTTF_{bn} = \int_0^\infty \left(2e^{-5\lambda t} - 5e^{-4\lambda t} + 2e^{-3\lambda t} + 2e^{-2\lambda t}\right)dt$$
$$= \frac{49}{60\lambda} \quad (3.43)$$

where

$MTTF_{bn}$ is the bridge network mean time to failure.

**Example 3.9**

Assume that an engineering system with five independent and identical units form a bridge network. Calculate the bridge network's reliability for a 200-hour mission and mean time to failure, if each unit's constant failure rate is 0.0004 failures per hour.

By substituting the specified data values into Equation (3.42), we get

$$R_{bn}(200) = 2e^{-5(0.0004)(200)} - 5e^{-4(0.0004)(200)} + 2e^{-3(0.0004)(200)} + 2e^{-2(0.0004)(200)}$$

$$= 0.9874$$

Similarly, by substituting the given data values into Equation (3.43), we obtain

$$MTTF_{bn} = \frac{49}{60(0.0004)}$$

$$= 2041.66 \text{ hours}$$

Thus, the bridge network's reliability and mean time to failure are 0.9874 and 2041.66 hours, respectively.

## 3.5 The importance, purpose, and results of maintainability efforts

The main reasons for the emphasis on maintainability are the alarmingly high operating and support costs of many systems and equipment, in part due to failures and necessary subsequent repairs. Some examples of these costs are the expense of maintenance personnel and their training, maintenance-related instructions and data, repair parts, test and support equipment, training-related equipment, and maintenance facilities.

The main objectives for applying maintainability engineering principles to engineering systems and equipment are as follows [8]:

- Reducing projected maintenance-related costs and time through design modifications directed at simplifications of maintenance.
- Using maintainability data for estimating item/system availability or unavailability.
- Determining labor hours and other related resources needed for performing the projected maintenance.

Past experiences indicate that when maintainability engineering principles are applied effectively to any system/product, the following results can be expected [9]:

- Reduced downtime for the system/product and consequently an increase in its operational readiness or availability.
- Maximizing operational readiness by eliminating failures that are due to wear-out or age.
- Efficient restoration of the system's/product's operating condition when random failures are the cause of downtime.

## 3.6 *Maintainability versus reliability*

Maintainability is a built-in design and installation characteristic that provides the resulting system/equipment/product with an inherent ability to be maintained, resulting in factors such as better mission availability and lower maintenance-related cost, required tools and equipment, required skill levels, and required man-hours.

In contrast, reliability is a design characteristic that results in durability of the equipment/system, as it conducts its assigned function according to a stated condition and time period. It is accomplished through actions such as choosing optimum engineering principles, testing, controlling-related processes, and satisfactory component/part sizing.

Eight of the important specific general principles of maintainability and reliability are presented below, separately for comparison purposes [10].

- **Specific general principles: Maintainability**
    - **I:** Reduce life cycle maintenance-related costs.
    - **II:** Reduce the amount, frequency, and complexity of maintenance-related tasks.
    - **III:** Lower mean time to repair (MTTR).
    - **IV:** Determine the extent of preventive maintenance to be conducted.
    - **V:** Provide for maximum interchangeability.
    - **VI:** Reduce the amount of supply needed.
    - **VII:** Reduce or eliminate altogether the need for maintenance.
    - **VIII:** Consider advantages of modular replacement versus repair or throwaway design.
- **Specific general principles: Reliability**
    - **I:** Maximize the use of standard parts/components.
    - **II:** Use fewer parts/components for conducting multiple functions.
    - **III:** Design for simplicity as much as possible.

- **IV:** Provide adequate safety factors between strength and peak stress values.
- **V:** Provide fail-safe designs as much as possible.
- **VI:** Provide redundancy when needed.
- **VII:** Minimize stress on parts and components as much as possible.
- **VIII:** Make use of parts and components with proven reliability.

## 3.7 Maintainability functions

Just like in other areas of engineering, probability distributions play an important role in maintainability engineering. They are used for representing repair times of systems, equipment, and parts. In this case, after the identification of the repair time distribution, the corresponding maintainability function may be obtained. This function is concerned with predicting the probability that a repair, beginning at time t = 0, will be accomplished in a time t.

Mathematically, the maintainability function is expressed by [8, 9]

$$MF(t) = \int_0^t f_r(t)dt \tag{3.44}$$

where

t is time.
$f_r(t)$ is the probability density function of the repair time.
MF(t) is the maintainability function.

Maintainability functions for exponential, Rayleigh, and Weibull probability distributions are obtained below [8–12].

### 3.7.1 Exponential distribution: Maintainability function

This distribution is simple and straightforward to handle and is quite useful for representing repair times. Its probability density function in regard to repair times is expressed by

$$f_{re}(t) = \mu e^{-\mu t} \tag{3.45}$$

where

t is the variable repair time.
μ is the constant repair rate or reciprocal of the MTTR.
$f_{re}(t)$ is the repair time probability density function of the exponential distribution.

By substituting Equation (3.45) into Equation (3.44), we obtain

$$MF_e(t) = \int_0^t \mu e^{-\mu t} dt \\ = 1 - e^{-\mu t} \tag{3.46}$$

where
$MF_e(t)$ is the maintainability function for exponential distribution.

Since $\mu = 1/MTTR$, Equation (3.46) becomes

$$MF_e(t) = 1 - e^{-\left(\frac{1}{MTTR}\right)t} \tag{3.47}$$

**Example 3.10**

Assume that the repair times of an engineering system are exponentially distributed with a mean value of 8 hours. Calculate the probability of completing a repair within 9 hours.

By substituting the given data values into Equation (3.47), we obtain

$$MF_e(9) = 1 - e^{-\left(\frac{1}{8}\right)(9)} \\ = 0.6753$$

Thus, the probability of completing the repair within 9 hours is 0.6753.

### 3.7.2 *Rayleigh distribution: Maintainability function*

This distribution is often used in reliability-related studies, and it can also be used for representing corrective maintenance times (i.e., repair times). Its probability density function in regard to repair times is expressed by

$$f_{rr}(t) = \frac{2}{\theta^2} t e^{-\left(\frac{t}{\theta}\right)^2} \tag{3.48}$$

where
t is the variable repair time.
$\theta$ is the distribution scale parameter.
$f_{rr}(t)$ is the repair time probability density function.

By inserting Equation (3.48) into Equation (3.44), we obtain

$$MF_r(t) = \int_0^t \left(\frac{2}{\theta^2}\right) t e^{-\left(\frac{t}{\theta}\right)^2} dt$$
$$= 1 - e^{-\left(\frac{t}{\theta}\right)^2} \tag{3.49}$$

where
$MF_r(t)$ is the maintainability function for Rayleigh distribution.

### 3.7.3 Weibull distribution: Maintainability function

Sometimes this distribution is used for representing repair times, particularly for electronic equipment. Its probability density function in regard to repair times is defined by

$$f_{rw}(t) = \frac{\alpha}{\theta^\alpha} t^{\alpha-1} e^{-\left(\frac{t}{\theta}\right)^\alpha} \tag{3.50}$$

where
t is the variable repair time.
$\alpha$ is the distribution shape parameter.
$\theta$ is the distribution scale parameter.
$f_{rw}$ is the repair time probability density function.

By substituting Equation (3.50) into Equation (3.44), we obtain

$$MF_w(t) = \int_0^t \frac{\alpha}{\theta^\alpha} t^{\alpha-1} e^{-\left(\frac{t}{\alpha}\right)^\alpha} dt$$
$$= 1 - e^{-\left(\frac{t}{\theta}\right)^\alpha} \tag{3.51}$$

where
$MF_w(t)$ is the maintainability function for Weibull distribution.

It is to be noted that at $\alpha = 1$ and $\alpha = 2$, Equation (3.51) reduces to Equations (3.46) and (3.49), respectively. Thus, exponential and Rayleigh distributions are the special case distributions of the Weibull distribution.

## 3.8 The role of engineers in regard to safety

Today, engineering systems/products have become highly sophisticated and complex. The matter of safety related to such systems/products has become a highly challenging issue for engineers who are pressured to complete new designs quickly and at lower costs due to competition and other factors. Past experiences over the years clearly indicate that this, in turn, generally results in more design-related deficiencies, errors, and causes of accidents. Design-related deficiencies, directly or indirectly, can contribute to or cause accidents.

As per Ref. [13], design-related deficiency may result because a designer/design:

- Fails to prescribe an effective operational procedure in conditions where a hazard might exist.
- Relies on users of product/item for avoiding an accident.
- Places an unreasonable level of stress on all potential users/operators.
- Creates an unsafe characteristic of a system/product/item.
- Creates an arrangement of operating controls and other devices that quite significantly increase reaction time during an emergency or is conducive to errors.
- Incorporates weak warning mechanisms rather than providing a safe design to get rid of hazards.
- Fails to eradicate or reduce the occurrence of human error.
- Fails to foresee an unexpected application of a system/product/item or its direct or indirect potential consequences.
- Does not appropriately determine or consider the consequences of an action, error, omission, or failure.
- Fails to properly warn of an expected potential hazard.
- Fails for providing an acceptable level of protection in a user's personal protective equipment.
- Violates general tendencies/capabilities of all potential users.
- Offers rather confusing, incorrect, or unfinished concepts/products.

## 3.9 *Safety management principles and organization tasks for product safety*

Over the years, professionals working in the safety area have developed many safety management principles. Nine of these principles are as follows [14–16]:

- **Principle I:** The key for the successful line-safety performance is management-related procedures that clearly factor in accountability.
- **Principle II:** The symptoms that indicate something is wrong in the management system include unsafe conditions, unsafe actions, and accidents.
- **Principle III:** The function of safety is to discover and define the operational errors that result in accidents.

- **Principle IV:** In developing an effective safety system, the three main subsystems that must be considered with utmost care are the managerial, the behavioral, and the physical subsystems.
- **Principle V:** Safety should be managed just like any other function in a company/organization. More clearly, management should direct safety by setting attainable goals and by planning, organizing, and controlling the attainment of such goals successfully.
- **Principle VI:** There is no single method for effectively achieving safety in an organization. But, for a safety system to be effective, it must satisfy system to be effective; it must satisfy certain criteria, such as involve worker participation, be flexible, have the top management visibly showing its full support to safety, and so on.
- **Principle VII:** Under most circumstances, usual human behavior is an unsafe behavior. Thus, it is the management team's responsibility as leaders for making appropriate changes to the environment that clearly fosters unsafe behavior in order to encourage safe behavior.
- **Principle VIII:** There are specific sets of circumstances that can predictably result in severe injuries. These circumstances are high-energy sources, non-productive activities, certain construction conditions, and abnormal non-routine activities.
- **Principle IX:** The causes that lead to unsafe behavior can be highlighted, classified, and controlled; the classification of the causes includes the employee's decision to err, traps, and overload.

An organization concerned with product safety carry out various types of tasks. Some of these tasks are presented below [14, 17]:

- Review government and non-government requirements directly or indirectly related to product safety.
- Review safety-related customer complaints and field reports.
- Review hazards and mishaps in existing similar products/items/systems for avoiding repetition of such hazards in future products/items/systems.
- Review product-test reports for determining shortcomings or trends in regard to safety.
- Review the product/system for establishing whether the potential hazards have been eradicated or controlled.
- Review warning labels—regarding safety factors such as satisfying requirements of, adequacy, and compatibility to warnings in the instruction manuals—that are to be placed on products/systems.
- Review all proposed system/product operation and maintenance documents with respect to safety.

- Participate in reviewing accident-related claims or recall actions by government agencies and recommend appropriate remedial actions for justifiable claims or recalls.
- Develop directives and programs related to product/system safety.
- Develop appropriate mechanisms by which the safety program can be monitored effectively.
- Develop appropriate safety criteria on the basis of applicable voluntary and governmental standards for use by the subcontractor, vendor, and company design professionals.
- Determine whether items such as warning and monitoring devices, protective equipment, or emergency equipment are necessary in handling or using the product.
- Provide appropriate assistance to designers in selecting alternative means for eradicating or controlling hazards or other safety-associated problems in initial designs.

## 3.10 *Product hazard classifications*

There are many product-related hazards and they may be grouped under the following six classifications [18]:

- **Classification I: Environmental hazards.** These hazards may be classified under two groups: internal hazards and external hazards. The internal hazards are associated with the changes in the surrounding environment that result in internal damage in the product/system. These hazards can be eliminated altogether or minimized by carefully considering, during the design phase, factors such as extremes of vibrations, ambient noise level, temperatures, electromagnetic radiation, and atmospheric contaminants. The external hazards are the hazards posed by system/product during its life span, and they include service-life operational hazards, maintenance hazards, and disposal hazards.
- **Classification II: Misuse-and-abuse hazards.** These hazards are concerned with product usage by humans. Past experiences over the years clearly indicate that product/system misuse can result in serious injuries. Also, product-abuse can result in hazardous situations or injuries; two examples of abuse are lack of proper maintenance and poor operating practices.
- **Classification III: Energy hazards.** These hazards may be classified under two groups: kinetic energy hazards and potential energy hazards. The kinetic energy hazards pertain to parts/items that have energy because of their motion, and some examples of such parts/items are fly wheels, fan blades, and loom shuttles. Any object that directly or indirectly interferes with their motion can experience extensive damage.

The potential energy hazards pertain to parts/items that store energy, and some examples of such parts/items are counterbalancing weights, springs, electronic capacitors, and compressed-gas receivers. During the equipment-servicing processes, such hazards are very important because stored energy can cause serious injury when released suddenly.

- **Classification IV: Human-factor hazards.** These hazards are associated with poor design with regard to humans, more clearly, to their intelligence, education, visual angle, physical strength, height, computational ability, length of reach, etc.
- **Classification V: Electrical hazards.** Two principal components of these hazards are shock and electrocution. Past experiences over the years indicate that the major electrical hazard to system/property/product stems from electrical faults, often referred to as short circuits.
- **Classification VI: Kinematic hazards.** These hazards are associated with circumstances when items/parts come together while moving and lead to crushing, cutting, or pinching any item/object caught between them.

## 3.11 *Accident causation theories*

There are many accident-causation theories and two such theories are as follows [19]:

### 3.11.1 *The 'Domino' accident-causation theory*

This theory is encapsulated in ten statements called the 'Axioms of Industrial Safety' by H.W. Heinrick [20]. All these ten axioms are presented below [14, 19, 20].

- **Axiom I:** An accident can take place only when someone commits an unsafe act or/and there is a physical or mechanical hazard.
- **Axiom II:** Most accidents are caused by the unsafe acts of people.
- **Axiom III:** Supervisors play a very important role in the prevention of industrial accidents.
- **Axiom IV:** The severity of an injury is largely fortuitous, and the specific accident that caused it is normally preventable.
- **Axiom V:** The injuries' occurrence results from a completed sequence of numerous factors, the last one of which is the accident itself.
- **Axiom VI:** The most useful accident-prevention approaches are analogous to the quality and productivity methods.

- **Axiom VII:** There are two types of costs (i.e., direct costs and indirect costs) of an accident. Some examples of the direct costs are liability claims, hospital-related expenses, and compensation.
- **Axiom VIII:** An unsafe act by an individual or an unsafe condition does not always immediately result in an accident/injury.
- **Axiom IX:** Management should always assume full responsibility for safety because it is in the best position for achieving the end results effectively.
- **Axiom X:** The reasons why people commit unsafe acts can be quite useful in selecting appropriate corrective measures.

As per Ref. [20], the five specific factors in the sequence of events leading up to an accident are as follows:

- **Factor I: Ancestry and social environment.** In this factor, it is assumed that negative character traits, such as stubbornness, avariciousness, and recklessness that might lead individuals for behaving in an unsafe manner, can be acquired as a result of the social environment or surroundings or inherited through one's ancestry.
- **Factor II: Fault of a person.** In this factor, it is assumed that negative character traits (i.e., whether inherited or acquired) such as nervousness, excitability, ignorance of safety-related practices, recklessness, and violent temper constitute proximate reasons for committing unsafe acts or for the existence of mechanical or physical hazards.
- **Factor III: Unsafe act/physical or mechanical hazard.** In this factor, it is assumed that unsafe acts by individuals (i.e., starting machinery/equipment without warning, removing safeguards, standing under suspended loads) and physical or mechanical hazards (i.e., inadequate light, unguarded point of operation, absence of guardrails, unguarded gears) are the very direct causes for the accidents' occurrences.
- **Factor IV: Accident.** In this factor, it is assumed that events such as striking of humans by flying objects and falls of humans are the typical examples of accidents that cause injury.
- **Factor V: Injury.** In this factor, it is assumed that the typical injuries that directly result from accidents include lacerations and fractures.

### 3.11.2 *The 'human factors' accident-causation theory*

This theory is based on the assumption that accidents occur due to a chain of events directly or indirectly linked to human error. It consists of three main factors, shown in Fig. 3.7, that lead to human error's occurrence [19, 21].

These three main factors are inappropriate activities, overload, and inappropriate response/incompatibility. The factor 'inappropriate activities'

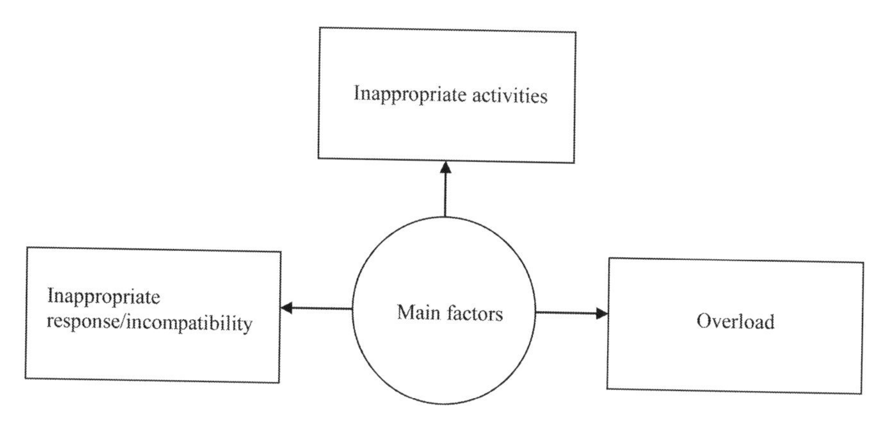


*Figure 3.7* Three main factors that lead to the human error's occurrence.

is concerned with inappropriate activities undertaken due to human error, poor judgement regarding the degree of risk involved in a given task, and subsequently acting on that misjudgment.

The factor 'overload' is concerned with an imbalance between an individual's capacity at any time and the lead he/she is carrying in a given state. The capacity of a person is the product of factors such as stress, fatigue, natural ability, the degree of training, physical condition, and state of mind. The overload carried by an individual is composed of tasks for which he/she has responsibility, along with additional burdens resulting from the situational factors (i.e., unclear instructions, risk level, etc.), internal factors (i.e., worry, emotional stress, personal problems, etc.), and environmental factors (i.e., noise, distractions, etc.)

Finally, the factor 'inappropriate response/incompatibility' is an important factor for the human error's occurrence, and three examples of inappropriate response by a person are as follows [14, 21]:

**Example I:** A person removes a safeguard from a machine for increasing output.
**Example II:** A person detects in hazardous condition but takes no corrective measure.
**Example III:** A person totally disregards the specified safety procedures.

## 3.12 Problems

1. What are the phases of a bathtub hazard rate curve? Discuss the causes for the occurrence of failures in each phase?
2. Write down general equations for reliability and hazard rate functions.

3. Write down three different formulas for obtaining mean time to failure.
4. Prove Equation (3.11) with the aid of Equation (3.6).
5. Assume that an engineering system is composed of three independent, identical, and active units. At least one of the units must operate normally for the engineering system to operate successfully. The constant failure rate of a unit is 0.0008 failures per hour.

   Calculate the engineering system reliability for a 300-hour mission and mean time to failure.
6. What are the main objectives for applying maintainability engineering principles to engineering systems and equipment?
7. Compare maintainability with reliability.
8. Write down general equation for maintainability function.
9. Write down at least nine safety management principles.
10. Describe the 'Domino' accident-causation theory.

## References


1. Dhillon, B.S., Design Reliability: Fundamentals and Applications, CRC Press, Boca Raton, Florida, 1999.
2. Kapur, K.C., Reliability and Maintainability, in Handbook of Industrial Engineering, edited by G. Salvendy, John Wiley and Sons, New York, 1982, pp. 8.5.1–8.5.34.
3. Dhillon, B.S., Life Distributions, IEEE Transactions on Reliability, Vol. 30, No. 5, 1981, pp. 457–460.
4. Shooman, M.L., Probabilistic Reliability: An Engineering Approach, McGraw-Hill, New York, 1968.
5. Dhillon, B.S., Reliability, Quality, and Safety for Engineers, CRC Press, Boca Raton, Florida, 2005.
6. Sandler, G.H., System Reliability Engineering, Prentice Hall, Englewood Cliffs, New Jersey, 1963.
7. Lipp, J.P., Topology of Switching Elements versus Reliability, Transactions on IRE Reliability and Quality Control, Vol. 7, 1957, pp. 21–34.
8. Dhillon, B.S., Engineering Maintainability, Gulf Publishing, Houston, Texas, 1999.
9. AMCP 706-133, Engineering Design Handbook: Maintainability Engineering Theory and Practice, Department of Defense, Washington, D.C., 1976.
10. AMCP 706-134, Engineering Design Handbook: Maintainability Guide for Design, Department of Defense, Washington, D.C., 1972.
11. Dhillon, B.S., Reliability Engineering in Systems Design and Operation, Van Nostrand Reinhold, New York, 1983.
12. Blanchard, B.S., Verma, D., Peterson, E.L., Maintainability, John Wiley and Sons, New York, 1995.
13. Hammer, W., Price, D., Occupational Safety Management and Engineering, Prentice Hall, Upper Saddle River, New Jersey, 2001.
14. Dhillon, B.S., Engineering Safety: Fundamentals, Techniques, and Applications, World Scientific Publishing, River Edge, New Jersey, 2003.

15. Petersen, D., Safety Management, American Society of Safety Engineers, Des Plaines, Illinois, 1998.
16. Petersen, D., Techniques of Safety Management, McGraw-Hill, New York, 1971.
17. Hammer, W., Product Safety Management and Engineering, Prentice Hall, Englewood Cliffs, New Jersey, 1980.
18. Hunter, T.A., Engineering Design for Safety, McGraw-Hill, New York, 1992.
19. Goetsch, D.L., Occupational Safety and Health, Prentice Hall, Englewood Cliffs, New Jersey, 1996.
20. Heinrich, H.W., Industrial Accident Prevention, McGraw-Hill, New York, 1959.
21. Heinrich, H.W., Petersen, D., Roos, N., Industrial Accident Prevention, McGraw-Hill, New York, 1980.

*chapter four*

# *Methods for performing reliability, maintainability, and safety analysis*

## 4.1 Introduction

Over the years, a large amount of published literature in the areas of reliability, maintainability, and safety have appeared in the form of books, technical reports, journal articles, and conference proceedings articles [1–8]. Many of these publications report the development of various types of methods and approaches for performing reliability, maintainability, and safety analyses. Some of these methods and approaches can be used quite effectively for performing analysis in all these three areas. The others are more confined to a specific area (i.e., reliability, safety, or maintainability).

An example of the methods and approaches that can be used to perform analysis in reliability, maintainability, and safety areas is the fault tree analysis (FTA). The FTA method was developed in the early 1960s for analyzing the safety of rocket launch control systems in the United States. Today, FTA is being used in many diverse areas for analyzing various types of problems.

This chapter presents a number of methods considered useful for performing reliability, maintainability, and safety analysis of engineering systems.

## 4.2 Fault tree analysis (FTA)

This method was developed in the early 1960s at the Bell Laboratories for evaluating the safety and reliability of the Minuteman Launch Control System [9]. FTA starts by defining a system's undesirable state (event) and then analyzes the system for determining all possible situations that can lead to the occurrence of the undesirable event. Thus, it highlights all possible failure causes at all possible levels associated with a system under consideration as well as the relationship between causes. FTA can be used to analyze various types of reliability, maintainability, and safety-related problems. Nonetheless, the main objectives of performing FTA are as follows [1, 3]:

- To comprehend the functional relationships of system failures.
- To verify the system's ability to meet its imposed safety-related requirements.
- To understand the degree of protection that the design concept provides against the occurrence of failures.
- To meet jurisdictional-related requirements.
- To highlight cost-effective improvements and critical areas.

There are many prerequisites associated with FTA, the six main ones are as follows [1]:

- Clearly defined analysis objectives and scope.
- Clear identification of all related assumptions.
- Clear definition of what constitutes system/item failure: the undesirable event.
- Clear understanding of design, operation, and maintenance aspects of system/item under consideration.
- Clearly defined system/item interfaces and system/item physical bounds.
- A comprehensive review of system/item operational experience.

FTA uses various types of symbols, and four commonly used symbols in fault tree construction are shown in Fig. 4.1 [9]. The circle denotes a basic fault event or the failure of an elementary part/component. The event's occurrence probability and failure and repair rates are generally obtained from empirical data. The rectangle denotes a fault event that results from the combination of fault events through the input of a logic gate. The AND gate denotes that an output fault event occurs only if all the input fault events occur. Finally, the OR gate denotes that an output fault event occurs if one or more of the input fault events occur.

The probabilities of the occurrence of the output fault events of logic gates AND and OR are given by

**AND gate**

$$P(X_0)=\prod_{i=1}^{m}P(X_i) \tag{4.1}$$

where

$P(X_0)$ is the occurrence probability of the AND gate output fault event $X_0$.

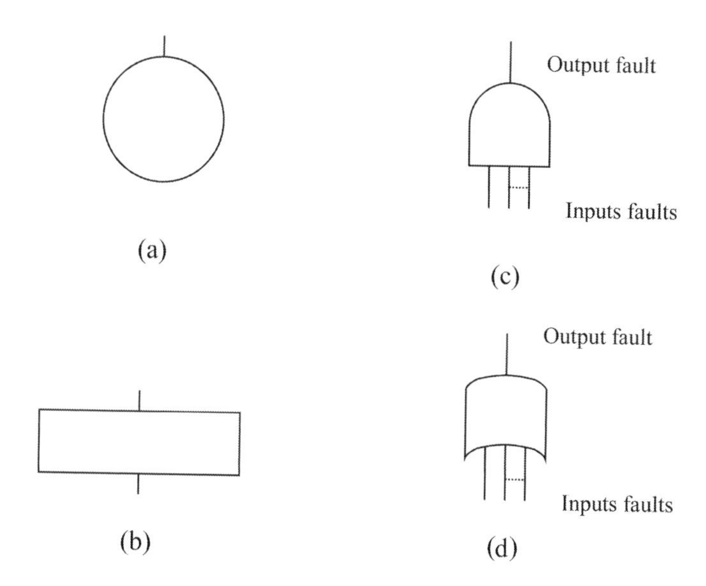


***Figure 4.1*** Four commonly used fault tree symbols: (a) circle, (b) rectangle, (c) AND gate, and (d) OR gate.

$P(X_i)$ is the occurrence probability of input fault event $X_i$, for i = 1, 2, 3, . . . , m.
m is the number of independent input fault events.

**OR gate**

$$P(E_0) = 1 - \prod_{i=1}^{m} \{1 - P(E_i)\} \tag{4.2}$$

where

$P(E_0)$ is the occurrence probability of the OR gate output fault event $E_0$.
$P(E_i)$ is the occurrence probability of input fault event $E_i$, for i = 1, 2, 3, . . . , m.

The following example demonstrates, the application of FTA to a reliability-related problem:

**Example 4.1**

Assume that a windowless room contains three light bulbs and one switch. Develop a fault tree for the top (undesired) fault event, dark room, if the switch can only fail to close.

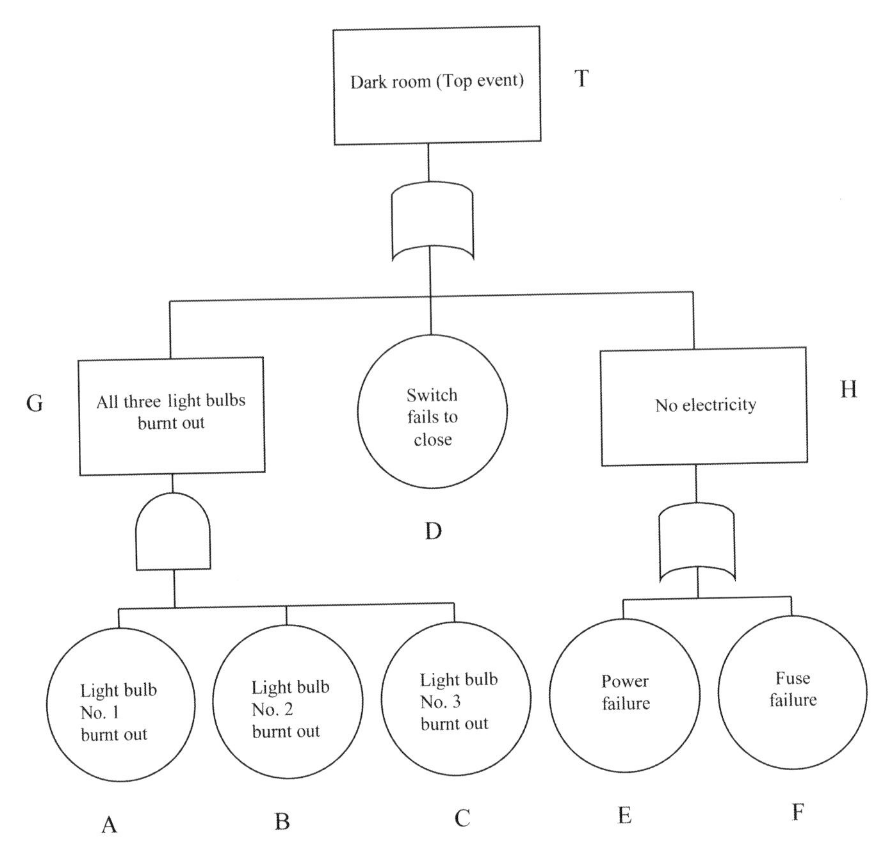


***Figure* 4.2** A fault tree for the top fault event: dark room.

In this case, there can be no light in the room (i.e., dark room) if the switch fails to close, if all the light bulbs burn out, or if there is no incoming electricity. Furthermore, there can only be no incoming electricity if there is a fuse failure or power failure.

Using the symbols shown in Fig. 4.1, a fault tree for the example is shown in Fig. 4.2. It is to be noted that the single capital letters in Fig. 4.2 diagram denote corresponding fault events [e.g., T: Dark room (top fault event)].

**Example 4.2**

Assume that the probabilities of occurrence of independent fault events A, B, C, D, E, and F in Fig. 4.2 are 0.09, 0.08, 0.07, 0.01, 0.02, and 0.04, respectively. Calculate the probability of occurrence of the top fault event T (dark room) by using Equations (4.1) and (4.2).

By substituting the specified occurrence probability values of fault events A, B, and C into Equation (4.1), we obtain

$$P(G) = (0.09)(0.08)(0.07) = 0.000504$$

where

P(G) is the probability of occurrence of fault event G (all three light bulbs burnt out).

Similarly, by substituting the specified occurrence probability values of fault events E and F into Equation (4.2), we obtain

$$P(H) = 1 - [(1 - 0.02)(1 - 0.04)]$$
$$= 0.0592$$

where

P(H) is the probability of occurrence of fault event H (no electricity).

By substituting the given data value and the above two calculated values into Equation (4.2), we get

$$P(T) = 1 - [(1 - 0.000504)(1 - 0.01)(1 - 0.0592)]$$
$$= 0.06907$$

where

P(T) is the probability of occurrence of fault event T (dark room).

Thus, probability of occurrence of the top fault event T (dark room) is 0.06907.

## 4.3 *Failure modes and effect analysis (FMEA)*

This is a widely used method for analyzing engineering systems in regard to reliability, and it may simply be described as an approach for conducting analysis of each system failure mode to examine its effects on the whole system [10]. When the FMEA is extended for categorizing the effect of each potential failure according to its severity, the method is known as failure mode effects and criticality analysis (FMECA) [1].

The history of FMEA may be traced back to the early years of 1950s with the development of flight control systems when the U.S. Navy's Bureau of Aeronautics, in order to establish a mechanism for reliability control over the systems' design effort, developed a requirement called 'Failure Analysis' [11]. Subsequently, failure analysis became known as failure effect analysis and then FMEA [12]. The National Aeronautics and Space Administration (NASA) extended FMEA for categorizing each potential failure's effect according to its severity and called it FMECA [13]. Subsequently, the U.S. Department of Defense, in 1980, developed a military standard entitled 'Procedures for Performing a Failure Mode, Effects, and Criticality Analysis' [14].

The following seven steps are involved in performing FMEA [9]:

- **Step 1:** Define system boundaries and related requirements in detail.
- **Step 2:** List all system parts/components and subsystems.
- **Step 3:** List all possible failure modes and describe and highlight the part/component under consideration.
- **Step 4:** Assign appropriate failure rate or probability to each part/component failure mode.
- **Step 5:** List each failure mode's effects on subsystems and the system/plant.
- **Step 6:** Enter appropriate remarks for each failure mode.
- **Step 7:** Review each critical failure mode and take appropriate action.

Additional information on this method is available in Refs. [1, 15].

## 4.4 *Markov method*

This is one of the most widely used methods to perform various types of reliability, maintainability, and safety analysis and is named after a Russian mathematician Andrei Andreyevich Markov (1856–1922). In the area of reliability, it is mainly used to analyze repairable and non-repairable systems with constant failure/repair rates. The method is subject to the following assumptions [16]:

- The transitional probability from one state to another in the finite time interval $\Delta t$ is given by $\lambda \Delta t$, where $\lambda$ is the constant transition rate (e.g., the failure or repair rate) from one system state to another.
- The probability of occurrence of more than one transition in finite time interval $\Delta t$ from one state to another is very small or negligible (e.g., $[\lambda \Delta t][\lambda \Delta t] \rightarrow 0$).
- All occurrences are independent of each other.

The following example demonstrates the application of this method to an engineering system [1]:

**Example 4.3**

Assume that an engineering system can be in one of two states: operating normally or failed. Its constant failure and repair rates are $\lambda_e$ and $\mu_e$, respectively. The engineering system state space diagram is

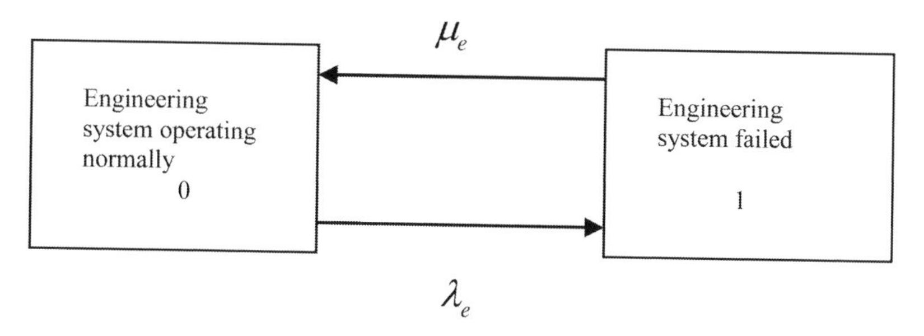


***Figure 4.3*** Engineering system state space diagram.

shown in Fig. 4.3. The numerals in rectangles denote the engineering system states.

Obtain expressions, with the aid of Markov method, for engineering system time-dependent and steady-state availabilities and unavailabilities.

Using Markov method, we write down the following equations for the state space diagram shown in Fig. 4.3:

$$P_0(t+\Delta t) = P_0(t)(1-\lambda_e\Delta t) + \mu_e\Delta t P_1(t) \tag{4.3}$$

$$P_1(t+\Delta t) = P_1(t)(1-\mu_e\Delta t) + \lambda_e\Delta t P_0(t) \tag{4.4}$$

where

$P_0(t+\Delta t)$ is the probability that the engineering system is in operating state 0 at time $(t+\Delta t)$.

$P_1(t+\Delta t)$ is the probability that the engineering system is in failed state 1 at time $(t+\Delta t)$.

$P_0(t)$ is the probability that the engineering system is in operating state 0 at time t.

$P_1(t)$ is the probability that the engineering system is in failed state 1 at time t.

$\lambda_e\Delta t$ is the probability of the engineering system failure in finite time interval $\Delta t$.

$\mu_e\Delta t$ is the probability of the engineering system repair in finite time interval $\Delta t$.

$(1-\lambda_e\Delta t)$ is the probability of no failure in finite time interval $\Delta t$.

$(1-\mu_e\Delta t)$ is the probability of no repair in finite time interval $\Delta t$.

In the limiting case, Equations (4.3) and (4.4) become

$$\frac{dP_0(t)}{dt} + \lambda_e P_0(t) = \mu_e P_1(t) \tag{4.5}$$

$$\frac{dP_1(t)}{dt} + \mu_e P_1(t) = \lambda_e P_0(t) \tag{4.6}$$

At time t = 0, $P_0(0) = 1$ and $P_1(0) = 0$.

By solving Equations (4.5) and (4.6), we obtain

$$P_0(t) = \frac{\mu_e}{(\lambda_e + \mu_e)} + \frac{\lambda_e}{(\lambda_e + \mu_e)} e^{-(\lambda_e + \mu_e)t} \tag{4.7}$$

$$P_1(t) = \frac{\lambda_e}{(\lambda_e + \mu_e)} - \frac{\lambda_e}{(\lambda_e + \mu_e)} e^{-(\lambda_e + \mu_e)t} \tag{4.8}$$

Thus, the engineering system time-dependent availability and unavailability, respectively, are

$$A_e(t) = P_0(t) = \frac{\mu_e}{(\lambda_e + \mu_e)} + \frac{\lambda_e}{(\lambda_e + \mu_e)} e^{-(\lambda_e + \mu_e)t} \tag{4.9}$$

and

$$UA_e(t) = P_1(t) = \frac{\lambda_e}{(\lambda_e + \mu_e)} - \frac{\lambda_e}{(\lambda_e + \mu_e)} e^{-(\lambda_e + \mu_e)t} \tag{4.10}$$

where
$A_e(t)$ is the engineering system availability at time t.
$UA_e(t)$ is the engineering system unavailability at time t.

By letting time t go to infinity in Equations (4.9) and (4.10), we get [1]

$$A_e = \lim_{t \to \infty} A_e(t) = \frac{\mu_e}{\lambda_e + \mu_e} \tag{4.11}$$

and

$$UA_e = \lim_{t \to \infty} UA_e(t) = \frac{\lambda_e}{\lambda_e + \mu_e} \tag{4.12}$$

where
$A_e$ is the engineering system steady state availability.
$UA_e$ is the engineering system steady state unavailability.

Since $\lambda_e = \frac{1}{MTTF_e}$ and $\mu_e = \frac{1}{MTTR_e}$, Equations (4.11) and (4.12) become

$$A = \frac{MTTF_e}{MTTF_e + MTTR_e} \tag{4.13}$$

and

$$UA_e = \frac{MTTR_e}{MTTF_e + MTTR_e} \tag{4.14}$$

where
$MTTF_e$ is the engineering system mean time to failure.
$MTTR_e$ is the engineering system mean time to repair.

**Example 4.4**

Assume that the mean time to failure of an engineering system is 2000 hours, and its mean time to repair is 50 hours. Calculate the engineering system steady state availability and unavailability.

By substituting the specified data values into Equations (4.13) and (4.14), we get

$$A_e = \frac{2000}{2000 + 50} = 0.9756$$

and

$$UA_e = \frac{50}{2000 + 50} = 0.0244$$

Thus, the engineering system steady state availability and unavailability are 0.9756 and 0.0244, respectively.

## 4.5 *Cause and effect diagram*

This is a deductive analysis approach that can be quite useful in reliability, maintainability, and safety areas. It is to be noted that in the published literature, this method is also known as a fishbone diagram because it resembles the skeleton of a fish, or an Ishikawa diagram, after its originator, K. Ishikawa of Japan [17]. A cause of effect diagram makes use of a graphic fishbone for depicting the cause and effect relationship between an undesired event and its all associated contributing causes.

The right side (i.e., the fish head or the box) of the diagram denotes the effect (i.e., the problem or the undesired event), and left of this, all possible causes of the problem are connected to the central fish spine. The basic five steps involved in developing a cause and effect diagram are as follows:

- **Step 1:** Establish a problem statement or identify the effect to be investigated.
- **Step 2:** Brainstorm for identifying all possible causes for the problem under study.
- **Step 3:** Group all major causes into categories and stratify them.
- **Step 4:** Develop the diagram by linking the causes under appropriate process steps and write down the effect or problem in the diagram box (i.e., the fish head) on the right side.
- **Step 5:** Refine all cause categories by asking questions such as 'What causes this?' and 'What is the reason for the existence of this condition?'.

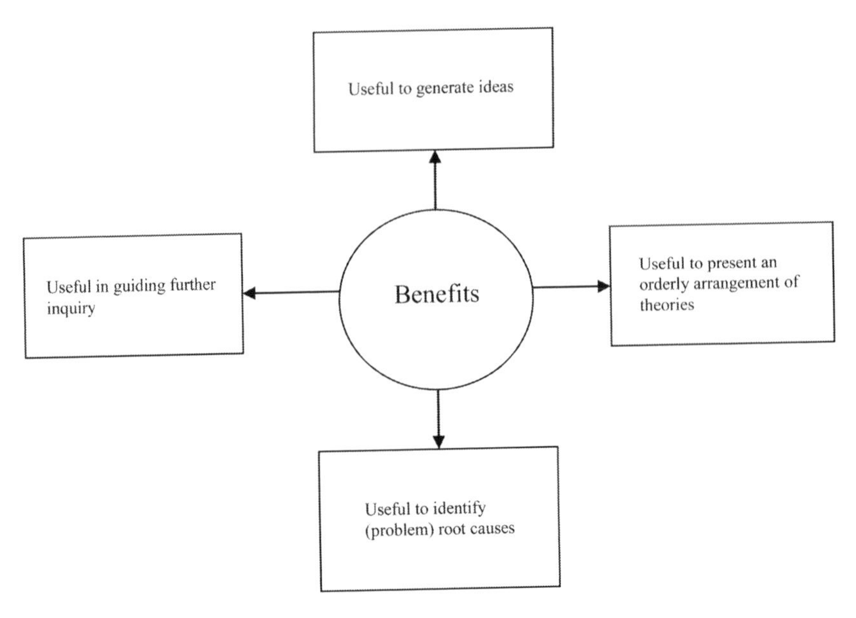


***Figure 4.4*** Important benefits of the cause and effect diagram.

Some of the important benefits of the cause and effect diagram are shown in Fig. 4.4.

Finally, it is added that a well-developed cause and effect diagram can be a very effective tool for identifying possible reliability, maintainability, and safety-related problems [2].

## *4.6 Probability tree analysis*

This method is used for performing task analysis by diagrammatically representing critical human-related actions and other events concerning the system under consideration. Often, this method is used for conducting task analysis in the technique for human error rate prediction [9, 18]. In this method, the diagrammatic task analysis is represented by the branches of the probability tree. More clearly, the branching limbs of the tree denote the outcome (success or failure) of each event, and each branch is assigned a value for the probability of occurrence.

Some of the benefits of this method are as follows [19, 20]:

- Lowers the error occurrence probabilities in computation because of simplification in the computational process.
- Incorporates, with some modifications, factors such as interaction effects, interaction stress, and emotional stress.
- A useful visibility tool.

The application of this method is demonstrated through the following two examples:

**Example 4.5**

Assume that an engineering technician conducts three independent safety-related tasks: x, y, and z. Each of these three tasks can be conducted either correctly or incorrectly, and task x is conducted before task y and task y before task z.

Develop a probability tree and obtain an expression for the probability of not successfully accomplishing the overall mission by the engineering technician.

In this case, the engineering technician first conducts task x correctly or incorrectly and then proceeds to conduct task y. Task y also can be conducted either correctly or incorrectly by the engineering technician. After task y, the technician proceeds to conduct task z. This task also can be conducted either correctly or incorrectly by the engineering technician. This entire scenario is shown in Fig. 4.5.

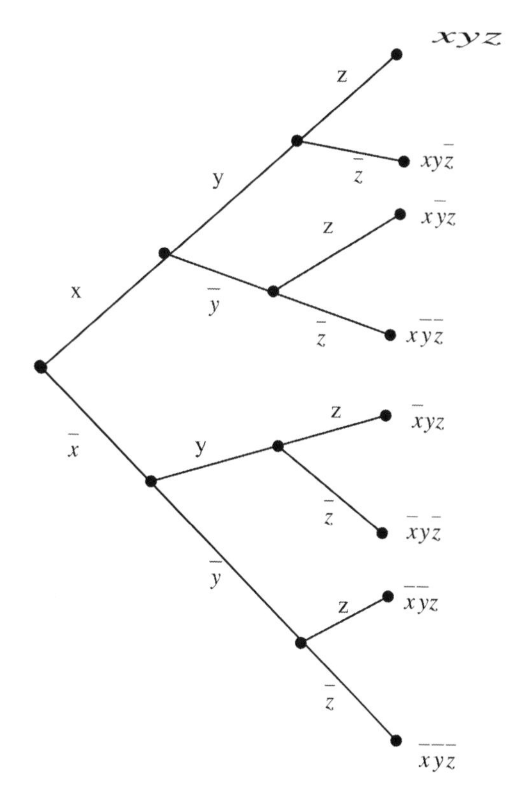


***Figure 4.5*** Probability tree diagram for the engineering technician conducting tasks x, y, and z.

Thus, the probability of not successfully accomplishing the overall mission by the engineering technician is expressed by

$$
\begin{aligned}
P_{ne} &= P(xy\bar{z}) + P(x\bar{y}z) + P(x\bar{y}\bar{z}) + P(\bar{x}yz) + P(\bar{x}y\bar{z}) + P(\bar{x}\bar{y}z) + P(\bar{x}\bar{y}\bar{z}) \\
&= P_x P_y P_{\bar{z}} + P_x P_{\bar{y}} P_z + P_x P_{\bar{y}} P_{\bar{z}} + P_{\bar{x}} P_y P_z + P_{\bar{x}} P_y P_{\bar{z}} + P_{\bar{x}} P_{\bar{y}} P_z + P_{\bar{x}} P_{\bar{y}} P_{\bar{z}}
\end{aligned}
\quad (4.15)
$$

where

$P_{ne}$ is the probability of not successfully accomplishing the overall mission by the engineering technician.

$P_x$ is the probability of conducting task x correctly by the engineering technician.

$P_y$ is the probability of conducting task y correctly by the engineering technician.

$P_z$ is the probability of conducting task z correctly by the engineering technician.

$P_{\bar{x}}$ is the probability of conducting task x incorrectly by the engineering technician.

$P_{\bar{y}}$ is the probability of conducting task y incorrectly by the engineering technician.

$P_{\bar{z}}$ is the probability of conducting task z incorrectly by the engineering technician.

Thus, Equation (4.15) is the expression for the probability of not successfully accomplishing the overall mission by the engineering technician.

**Example 4.6**

If the last task (i.e., task z) in Example 4.5 is eliminated, develop a probability tree and probability expression for successfully accomplishing the overall mission by the engineering technician. In addition, calculate the probability of not successfully accomplishing the overall mission by the engineering technician if the probabilities of conducting tasks x and y correctly are 0.7 and 0.9, respectively.

As there are only two tasks conducted by the engineering technician, the probability tree of Fig. 4.5 is reduced to the one shown in Fig. 4.6.

With the aid of Fig. 4.6 diagram, the probability of successfully accomplishing the overall mission by the engineering technician is

$$P_{se} = P(xy) = P_x P_y \quad (4.16)$$

where

$P_{se}$ is the probability of successfully accomplishing the overall mission by the engineering technician.

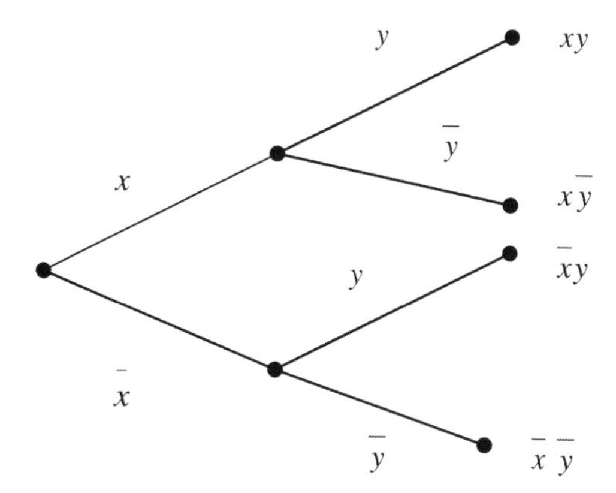


***Figure 4.6*** Probability tree for the engineering technician conducting tasks x and y only.

Similarly, with the aid of Fig. 4.6 diagram, the probability of not successfully accomplishing the overall mission by the engineering technician is

$$\begin{aligned} P_{ne} &= P(x\bar{y}) + P(\bar{x}y) + P(\overline{xy}) \\ &= P_x P_{\bar{y}} + P_{\bar{x}} P_y + P_{\bar{x}} P_{\bar{y}} \end{aligned} \tag{4.17}$$

where

$P_{ne}$ is the probability of not successfully accomplishing the overall mission by the engineering technician.

Because $P_{\bar{x}} = 1 - P_x$ and $P_{\bar{y}} = 1 - P_y$, Equation (4.17) reduces to

$$P_{ne} = 1 - P_x P_y \tag{4.18}$$

By inserting the specified data values into Equation (4.18), we obtain

$$\begin{aligned} P_{ne} &= 1 - (0.7)(0.9) \\ &= 0.37 \end{aligned}$$

Thus, the probability of not successfully accomplishing the overall mission by the engineering technician is 0.37.

## 4.7 Hazard and operability analysis (HAZOP)

This is a systematic and quite effective approach used for highlighting hazards and operating problems throughout a facility. In particular, past experiences over the years clearly indicate that it is extremely useful to highlight unforeseen hazards designed into facilities due to various reasons or introduced into existing facilities due to factors such as changes made to operating-related procedures or process conditions.

HAZOP has the following three basic objectives:

i. Produce a complete facility/process description.
ii. Review each and every facility/process part to find out how deviations from the design intentions can take place.
iii. Decide whether such deviations can lead to operating problems/hazards.

HAZOP study is conducted in five steps shown in Fig. 4.7 [3, 21].

Step 1 is an important step and is concerned with developing study objectives and scope by carefully considering all relevant factors. Step 2 is concerned with forming a HAZOP team by ensuring that its members are comprised of individuals from design and operation with proper experience to determine the effects of deviations from intended application.

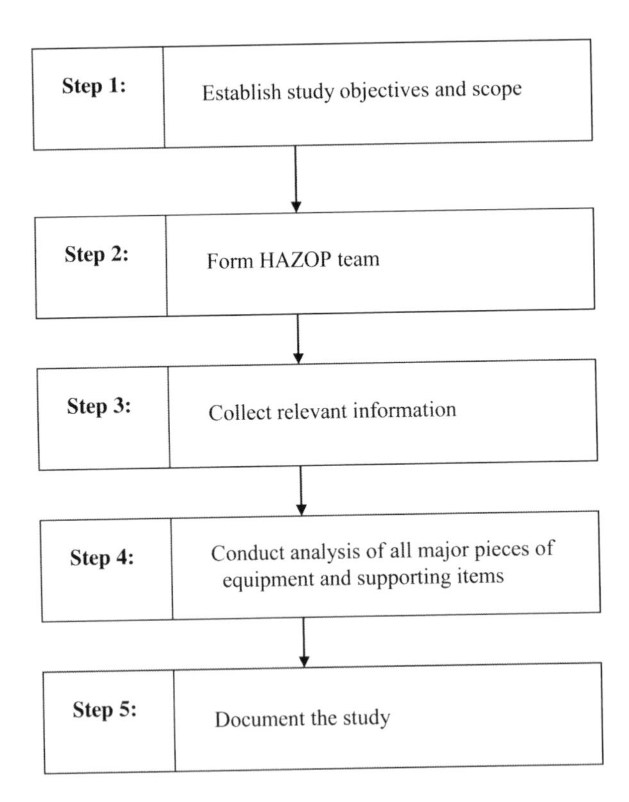


***Figure 4.7*** HAZOP study steps.

Step 3 is concerned with obtaining the necessary drawings, documentation, and process description. More clearly, this includes items such as layout drawings, equipment specifications, operating and maintenance procedures, process flow-sheets, process control logic diagrams, and emergency response procedures. Step 4 is concerned with conducting analysis of each and every major item of equipment, and all supporting equipment, instrumentation, and piping by utilizing the step 3 documents.

Finally, step 5 is basically concerned with documenting all the consequences of any deviation from the norm and a summary of deviation from the norm and a summary of deviations considered quite hazardous and credible.

## 4.8 *Technique of operations review (TOR)*

This method was developed in the early 1970s by D.A. Weaver of the American Society of Safety Engineers, and it seeks to highlight systemic causes rather than assigning blames in regard to safety [8]. TOR allows employees and management to work jointly for analyzing workplace failure, incidents, and accidents. More specifically, TOR may simply be described as a hands-on analytical methodology developed, for identifying the root system causes of an operation malfunction [8].

The method makes use of a worksheet containing simple terms that basically require 'yes/no' decisions. The TOR is activated by an incident occurring at a certain point in time and location involving certain persons. Furthermore, it may be stated that this method is not a hypothetical process and demands a systematic evaluation of the actual circumstances surrounding the incident. Ultimately, TOR leads to isolate the specific ways the organization/company failed to prevent, thus resulting in the accidents' occurrence.

TOR is composed of the following eights steps [6]:

- **Step 1:** Form the TOR team by carefully selecting its members from all the concerned areas.
- **Step 2:** Hold a roundtable session for departing common knowledge to all members of the team.
- **Step 3:** Highlight one key systemic factor that played a pivotal role in causing the incident/accident to occur. This factor must be based on consensus of the team and serves as an initial point for further investigation.
- **Step 4:** Make use of the team consensus for responding to a sequence of 'yes/no' options.
- **Step 5:** Evaluate the highlighted factors with care by ensuring the existence of the team consensus in regard to the evaluation of each factor.

- **Step 6:** Prioritize the contributory factors by starting with the most serious one.
- **Step 7:** Develop appropriate preventive/corrective strategies in regard to each contributory factor.
- **Step 8:** Conduct implementation of the strategies.

Finally, it is to be noted that just like in the case of any other method, TOR too has its strengths and weaknesses. The main strength of the TOR is the involvement of line personnel, and its main weakness is an after-the-fact process.

## 4.9 *Job safety analysis (JSA)*

This method is considered quite useful for uncovering and rectifying potential hazards that are intrinsic to or inherent in the workplace. Generally, safety professional, worker, supervisor, and management participate in JSA. JSA is conducted in five steps shown in Fig. 4.8 [3, 22].

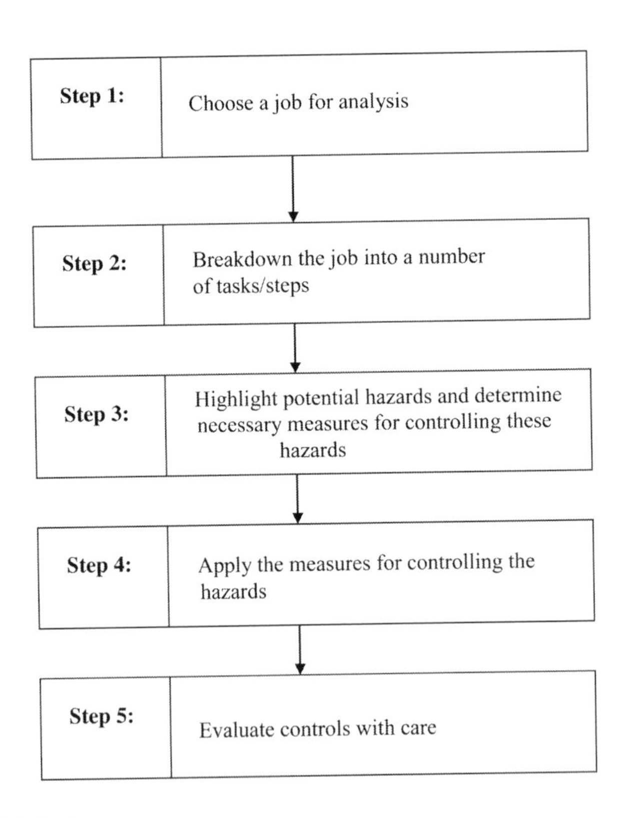


***Figure 4.8*** JSA steps.

Finally, it is to be noted that the degree of success very much depends on the rigor the JSA team exercises during the analysis process.

## 4.10 *Interface safety analysis (ISA)*

ISA is concerned with determining the incompatibilities between assemblies and subsystems of a product/item that could result in accidents. The analysis establishes that distinct parts/units can be integrated into a viable system, and a part's/unit's normal operation will not impair the performance or damage another part/unit or the whole system/product.

Although the analysis considers various relationships, they can be grouped under the following three areas [3, 23]:

- **Area I: Physical relationships.** These relationships are concerned with the physical aspects of items/systems. For example, two items/systems might be very well designed and manufactured and function quite well individually, but they fail to fit together effectively because of dimensional differences, or they may present other problems that directly or indirectly may lead to safety-related problems. Some examples of the other problems are: (i) impossible to tighten, join, or mate parts properly; (ii) impossible or restricted access to or egress from equipment; and (iii) a very small clearance between units, thus the units may be damaged during the removal or replacement process.
- **Area II: Flow relationships.** These relationships are concerned with two or more units/items. For example, the flow between two units/items may involve fuel, air, water, stream, lubricating oil, or electrical energy. Furthermore, the flow could be unconfined, such as heat radiation from one body to another. Normally, the frequent problems associated with many products are the proper flow of energy and fluids from one unit to another through confined passages, consequently, leading to direct or indirect safety-related problems. Nonetheless, some of the flow-related problem causes are faulty connections between units/parts, complete or partial interconnection failure, and so on. In the case of fluid, factors such as loss of pressure, flammability, lubricity, corrosiveness, contamination, odor, and toxicity must be considered seriously in regard to safety.
- **Area III: Functional relationships.** These relationships are concerned with multiple items/units. For example, in situations where outputs of an individual unit/item constitute the inputs to a downstream unit/item and in turn a safety hazard. The outputs could be in conditions such as zero outputs, excessive outputs, un-programmed outputs, degraded outputs, and erratic outputs.

## 4.11 Problems

1. Write an essay on FTA.
2. What are the main steps used for conducting FMEA?
3. What are the assumptions associated with the Markov method?
4. Assume that a windowless room contains two light bulbs and one switch. Develop a fault tree for the top (undesired) fault event, dark room, if the switch can only fail to close.
5. Assume that the constant failure rate of an engineering system is 0.005 failures/hour and its constant repair rate is 0.04 repairs/hour. Calculate the engineering system unavailability during a 150-hour mission period.
6. What are the important benefits of the cause and effect diagram?
7. Describe HAZOP.
8. Prove Equations (4.7) and (4.8) by using Equations (4.5) and (4.6).
9. Describe TOR.
10. What are the advantages of probability tree analysis?

## References

1. Dhillon, B.S., Design Reliability: Fundamentals and Applications, CRC Press, Boca Raton, Florida, 1999.
2. Dhillon, B.S., Engineering Maintainability, Gulf Publishing, Houston, Texas, 1999.
3. Dhillon, B.S., Engineering Safety: Fundamentals, Techniques and Applications, World Scientific Publishing, Upper Saddle River, New Jersey, 2003.
4. Dhillon, B.S., Engineering Systems Reliability, Safety, and Maintenance: An Integrated Approach, CRC Press, Boca Raton, Florida, 2017.
5. Dhillon, B.S., Failure Modes and Effects Analysis-Bibliography, Microelectronics, and Reliability, Vol. 32, 1992, pp. 719–731.
6. Goetsch, D.L., Occupational Safety and Health, Prentice Hall, Englewood Cliffs, New Jersey, 1996.
7. AMCP 706-133, Engineering Design Handbook: Maintainability Engineering Theory and Practice, Department of Defense, Washington, D.C., 1976.
8. Hallock, R.G., Technique of Operations Review Analysis: Determine Cause of Accident/Incident, Safety and Health, Vol. 60, No. 8, 1991, pp. 38–39.
9. Dhillon, B.S., Singh, C., Engineering Reliability: New Techniques and Applications, John Wiley and Sons, New York, 1981.
10. Omdahl, T.P., Editor, Reliability, Availability, and Maintainability (RAM) Dictionary, American Society for Quality Control (ASQC) Press, Milwaukee, WI, 1988.
11. MIL-F-18372 (Aer), General Specification for Design, Installation, and Test of Aircraft Flight Control Systems, Bureau of Naval Weapons, Department of the Navy, Washington, D.C., Para. 3.5.2.3.
12. Continho, J.S., Failure Effect Analysis, Transactions of the New York Academy of Sciences, Ser II, Vol. 26, 1963–1964, pp. 564–584.

13. Jordan, W.E., Failure Modes, Effects, and Criticality Analyses, Proceedings of the Annual Reliability and Maintainability Symposium, 1972, pp. 30–37.
14. MIL-STD-1629, Procedures for Performing a Failure Mode, Effects, and Criticality Analysis, Department of Defense, Washington, D.C., 1980.
15. Palady, P., Failure Modes and Effect Analysis, PT Publications, West Palm Beach, Florida, 1995.
16. Shooman, M.L., Probabilistic Reliability: An Engineering Approach, McGraw-Hill, New York, 1968.
17. Ishikawa, K., Guide to Quality Control, Asian Productivity Organization, Tokyo, 1976.
18. Dhillon, B.S., Human Reliability: With Human Factors, Pergamon Press, New York, 1986.
19. Swain, A.D., An Error-Cause Removal/Program for Industry, Human Factors, Vol. 12, 1973, pp. 207–221.
20. Dhillon, B.S., Robot System Reliability and Safety: A Modern Approach, CRC Press, Boca Raton, Florida, 2015.
21. CAN/CSA-Q6340-91, Risk Analysis Requirements and Guidelines, prepared by the Canadian Standards Association (CSA), 1991. Available from CSA, 178 Rexdale Boulevard, Rexdale, Ontario, Canada.
22. Hammer, W., Price, D., Occupational Safety Management and Engineering, Prentice Hall, Upper Saddle River, New Jersey, 2001.
23. Hammer, W., Product Safety Management and Engineering, Prentice Hall, Englewood Cliffs, New Jersey, 1980.

# chapter five

# Reliability management

## 5.1 Introduction

Nowadays, reliability management has become an important element of reliability engineering due to factors such as system complexity, sophistication, and size; cost and time constraints; and demanding reliability requirements. Reliability management is concerned with the direction and control of reliability activities of an organization. Some examples of these activities are developing reliability policies and goals, facilitating interactions of reliability workforce with other parts of the organization, and staffing.

The history of reliability management goes back to the late 1950s when the U.S. Air Force developed a reliability program management document (i.e., Exhibit 58-10) [1]. Subsequently, the U.S. Department of Defense's effort to develop requirements for an organized contractor reliability program resulted in the release of the military specification MIL-R-27542 [2]. Additional information on the history of reliability management is available in Ref. [3].

This chapter presents various important aspects of reliability management.

## 5.2 General management reliability program responsibilities and guiding force-related facts for the general management for an effective reliability program

In the success of a reliability program, general management plays an important role. Some of its responsibilities are as follows [3]:

- Developing appropriate reliability-related goals.
- Developing a mechanism for accessing information concerning the organization's current reliability performance in regard to its operations and products.
- Providing appropriate funds, manpower, authority, and scheduled time.

- Monitoring the program regularly and taking appropriate corrective measures in regard to associated procedures, organization, policies, and so on.
- Establishing an appropriate program for fulfilling set reliability goals/objectives and eradicating existing shortcomings.

Facts such as the following will be a good guiding force for the general management for having an effective reliability program [4]:

- Appropriate planned programs are required for application in design, manufacturing, testing, and field phases of the engineering product for controlling reliability.
- Reliability is a very important factor in the management, planning, and design of an engineering product.
- Reliability is established by the fundamental/basic design.
- In achieving the desired reliability in a mature engineering product in a timely manner, deficiency data collection, analysis, and feedback are very important.
- Any changes in maintenance, manufacturing, storage and shipping, testing, and usage in the field of the engineering product tend to decrease the reliability of the design.
- Improvement in reliability can be through design-related changes only.
- It is during the initial phases of the design and evaluation testing programs when high levels of reliability can be achieved most economically.
- Human error lowers the reliability of the design.

## 5.3 A procedure for developing reliability goals and useful guidelines for developing reliability programs

Fifteen steps of a procedure considered quite useful for developing reliability goals are as follows [5]:

- **Step 1:** Examine and clarify all preestablished requirements.
- **Step 2:** Examine the organizational objective as well as the organizational unit's mission.
- **Step 3:** Highlight desired important result areas.
- **Step 4:** Highlight the most promising payoff areas.
- **Step 5:** Select most promising result areas for pursuance.
- **Step 6:** Select goal candidates.
- **Step 7:** Review resource requirements to pursue each candidate goal for achieving it successfully.

- **Step 8:** Highlight anticipated problem areas in achieving goals.
- **Step 9:** Rank all candidate goals with respect to ease of payment and the degree of payoff.
- **Step 10:** Review goal interdependencies and make necessary adjustments in goal candidates for maximum coordination.
- **Step 11:** Review goals with respect to factors such as acceptability, attainability, compatibility, and measurability.
- **Step 12:** Make final selection of goals and establish appropriate milestones for their successful achievement.
- **Step 13:** Develop action plans for goal achievement.
- **Step 14:** Communicate goals in written form to all concerned people and review their progress periodically.
- **Step 15:** Make adjustments as appropriate.

In the published literature, for developing reliability programs, various guidelines have appeared. Twelve guidelines for developing reliability programs that appeared in the military specification MIL-R-27542 are as follows [6]:

- Conduct specification review.
- Assign reliability-associated goals for system under consideration.
- Conduct procedure and design reviews.
- Evaluate reliability margin.
- Develop a closed-loop system for failure reporting, analysis, and feedback to involved engineering personnel for corrective actions to stop re-occurrence.
- Develop an appropriate testing program.
- Assign responsibility for reliability to a single group.
- Establish an on-the-job training facility and program.
- Review reliability-related changes in specifications and drawings.
- Ensure the reporting of the reliability group to an appropriate authority so that the reliability-related work can be carried out effectively.
- Develop and maintain control during the production process through actions such as sample testing and inspection.
- Put maximum effort during the design phase for achieving maximum 'inherent equipment reliability'.

## 5.4 Reliability and maintainability management-related tasks in the product life cycle

To obtain the desired level of reliability of a product/system in the field, a series of management tasks in regard to reliability and maintainability must be conducted throughout the system life cycle. The system life cycle may be divided into four phases, as shown in Fig. 5.1 [7].

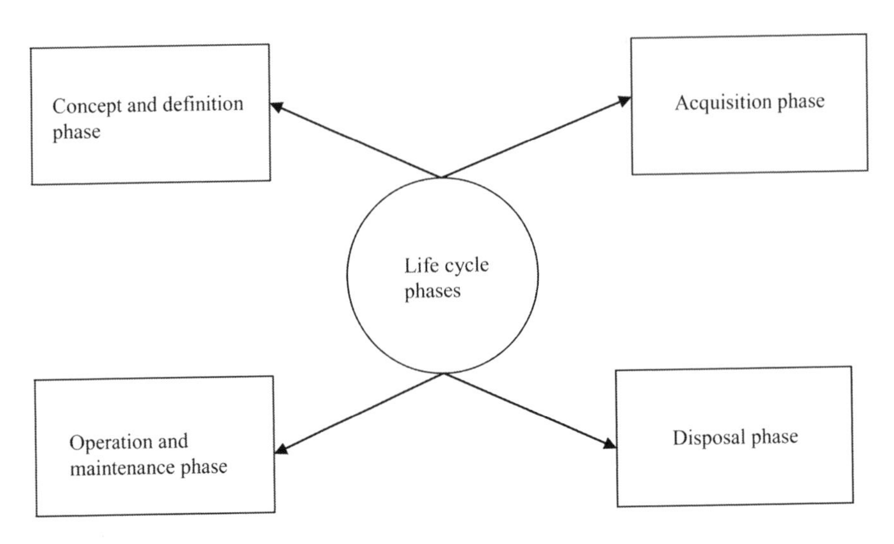


***Figure 5.1*** System/product life cycle phases.

The reliability and maintainability management-related tasks involved in each of these four phases are as follows:

### 5.4.1 *Concept and definition phase*

In this phase, system-related requirements are established and the basic characteristics are defined. During this phase, various reliability- and maintainability-related management tasks are conducted. Some of them are as follows:

- Defining system safety requirements.
- Defining a failure.
- Defining the basic maintenance philosophy.
- Defining the system capability requirements, management controls, part control requirements, and terms used.
- Defining the system reliability and maintainability goals and quantitative terms.
- Defining methods to be used during the design and manufacturing phase.
- Defining system environmental factors during its life cycle.
- Defining hardware and software standard documents to be used for fulfilling reliability and maintainability requirements.
- Defining constraints proven to be harmful to reliability.
- Defining the management controls needed for documentation.

### 5.4.2 *Acquisition phase*

This phase is concerned with activities associated with system acquisition and installation as well as planning for eventual support of the system. In this phase, there are many reliability and maintainability management-associated tasks. Some of these tasks are as follows:

- Define all the system technical requirements, design and development methods to be employed, type of evaluation methods to be used for assessing the system, demonstration requirements, and documents required as part of the final system.
- Define the kind of reviews to be conducted.
- Define all the reliability and maintainability requirements that must be satisfied.
- Define the type of field studies, if any, to be performed.
- Define the meaning of a degradation or a failure.
- Define the kind of logistics support needed.
- Define the cost-related restraints and the life cycle cost information to be developed.
- Define the type of data to be supplied by the manufacturer to the customer.

### 5.4.3 *Operation and maintenance phase*

This phase is concerned with tasks related to the maintenance activity, management of the engineering, and the support of the system over its entire operational life. Some of the reliability- and maintainability-related management tasks involved during this phase are as follows:

- Providing adequate maintenance tools and test equipment.
- Establishing failure data banks.
- Providing adequate and appropriately trained manpower.
- Preparing maintenance-related documents.
- Managing and predicting spare parts.
- Reviewing the documentation with respect to any engineering change.
- Collecting, monitoring, and analyzing reliability and maintainability data.
- Developing engineering change proposals.

### 5.4.4 *Disposal phase*

This phase is concerned with tasks that are needed for removing the system and its nonessential supporting material. Two of the reliability and maintainability management-associated tasks involved in this phase are estimating the final system life cycle cost and the reliability and maintainability values. The resulting life cycle cost takes into consideration the disposal action income or cost.

The final reliability and maintainability values are computed for the buyer of the used system as well as for use in purchasing of similar systems.

## 5.5 *Reliability management documents and tools*

Reliability management makes use of a variety of documents and tools. Some examples of these documents are in-house reliability manual, documents explaining policy and procedures, instructions and plans, international standards and specifications, and reports and drawings [7]. Similarly, some of the tools used by reliability management are critical path method (CPM), value engineering, and configuration management.

Some of the above items are described below.

### 5.5.1 *Reliability manual*

This is the backbone of any reliability organization. Its existence is very important for any organization irrespective of its size. A typical reliability manual covers items such as follows:

- Organizational responsibilities and structure.
- Product design phase-related procedures with respect to reliability.
- Failure data collection and analysis procedures and methods to be followed.
- Relationship with customers and suppliers.
- Reliability test and demonstration procedures and approaches.
- Company-wide reliability policy.
- Effective reliability methods, models, and so on.

### 5.5.2 *Value engineering*

This is a systematic and creative technique used for accomplishing a required function at the minimum cost [8]. Historical records indicate that the application of this technique has returned between $15 and $30 for each dollar spent [9].

There are many areas in which value engineering is useful including the following:

- Generating new ideas to solve problems.
- Highlighting areas requiring attention and improvement.

- Servicing as a vehicle for dialogue.
- Increasing the value of goods and services.
- Prioritizing.
- Assigning dollars on high-value items.
- Determining and quantifying intangibles.

Additional information on value engineering is available in Ref. [10].

### 5.5.3 Configuration management

During the development of an engineering system, many changes may take place; these changes may be concerned with system performance, size, weight, and so on. In such situation, configuration management is considered a useful tool for assuring the customer and the manufacturer that the end system/product will fully satisfy the contract specification. Thus, configuration management may be defined as the management of technical requirements which defines the engineering system/product as well as changes thereto.

The history of configuration management goes back to 1962, when the U.S. Air Force released a document entitled 'Configuration Management During the Development and Acquisition Phases', AFSCM 375-1 [9]. Nowadays, configuration management is well known in the industrial sector, and its benefits include formal establishment of objectives, reduction in overall cost, elimination of redundant efforts, precisely identified final system/product, facilitation of accurate data retrieval, and effective channeling of resources.

Additional information on configuration management is available in Ref. [11].

### 5.5.4 Critical path method (CPM)

This method along with the program evaluation and review technique (PERT) is often used for planning and controlling projects. It was developed in 1956 by E.I. DuPont de Nemours and company for scheduling design- and construction-associated activities [12]. The following six general steps are associated with CPM:

- **Step I:** Break down the project under consideration into individual tasks/jobs.
- **Step II:** Arrange the tasks/jobs into a logical network.
- **Step III:** Estimate the duration time of each task/job.
- **Step IV:** Develop a schedule.

- **Step V:** Highlight tasks/jobs that control the completion of the project.
- **Step VI:** Redistribute resources and funds for improving the schedule.

There are many advantages of CPM. Some of these are: it identifies critical work activities for completing the project on time, determines project duration systematically, determines the need for labor and resources in advance, shows interrelationships in work flow, and improves communication and understanding.

Additional information on CPM is available in Ref. [13].

## 5.6 *Reliability engineering department responsibilities and a reliability engineer's tasks*

There are various kinds of responsibilities of a reliability engineering department. Some of these are developing reliability policy, plans, and procedures; conducting reliability allocation and prediction; providing reliability-related inputs to design specifications and proposals; auditing reliability activities; conducting reliability demonstration and failure data collection and reporting; training reliability manpower; performing reliability-related research; monitoring reliability growth and the reliability activities of subcontractors (if any); conducting specification and design reviews with respect to reliability; performing failure data analysis; and consulting on reliability matters [2].

A reliability engineer performs various types of tasks during planning, design and development, manufacturing, and improvement phases of a system. Some of these tasks are as follows [2]:

- Participating in evaluating requests for proposals.
- Conducting analysis of a proposed design.
- Determining reliability of alternative designs.
- Participating in design reviews.
- Providing relevant reliability information to management.
- Analyzing customer complaints with reliability.
- Budgeting the tolerable system failure down to the component level.
- Running tests on the system, subsystems, and parts.
- Monitoring subcontractor's reliability performance.
- Developing reliability prediction models and techniques.
- Developing a reliability program plan.
- Investigating field failures.
- Securing resources for an effective reliability program during the planning stage.

## 5.7 *Pitfalls in reliability program management and useful rules for reliability professionals*

Past experiences indicate that many reliability program-related problems and uncertainties are due to pitfalls in reliability program management. These pitfalls are associated with areas such as follows [14]:

- **Reliability organization:** Two examples of reliability organization pitfalls are having several organizational tiers and a number of individuals with authority for making commitments without internal dialogue and coordination.
- **Reliability testing:** An example of the reliability testing pitfalls is the delayed start of the reliability demonstration test. Because, this type of delay may compound the problem of incorporating reliability-related changes into deliverable product/system to the customer.
- **Programming:** An example of the programming pitfalls is the assumption that each concerned person with the program understands the specified reliability requirements.
- **Manufacturing:** An example of the manufacturing phase pitfalls is the authorization of substitute parts, without paying proper attention to their effect on reliability, by parts buyers or others when the parts acquisition lead time is incompatible with the system/product manufacturing schedule.

For the effective implementation of reliability programs in an organization by the reliability manager or engineer, there are many useful rules. Some of these are as follows [15]:

- Learn to believe that new reliability methods are not the end result but are only the processes that permit one for achieving product and business goals.
- Ensure that all reliability functions are included in the reliability program plan properly.
- Develop in yourself clearly the belief that the reliability engineer is a pessimist, the designer is an optimist, and the combination is a success.
- Clearly develop for yourself how to be just as comfortable reporting successes as reporting failures in regard to reliability.
- When communicating with other individuals, make sure that you converse in a language they understand clearly. More clearly, avoid using terms and abbreviations with which they are uncomfortable.
- Avoid assuming that the reliability function is responsible for achieving reliability-related goals.
- Avoid making use of statistical jargon as much as possible especially when dealing with the top-level management personnel.

## 5.8 Problems

1. Discuss the general management reliability program-related responsibilities.
2. Write an essay on reliability management.
3. What are the facts that will be a good guiding force for the general management for having an effective reliability program?
4. Discuss the procedure for developing reliability goals.
5. What are the guidelines for developing reliability programs?
6. What are the reliability and maintainability management-related tasks in the product acquisition phase?
7. Discuss the items covered in the reliability manual.
8. Discuss the following items:
   - Value engineering
   - Configuration management
9. Discuss responsibilities of a reliability engineering department.
10. What are the tasks of a reliability engineer?

## References

1. Austin-Davis, W., Reliability Management: A Challenge, IEEE Transactions on Reliability, Vol. 12, 1963, pp. 6–9.
2. Dhillon, B.S., Engineering Reliability Management, IEEE Journal on Selected Areas in Communications, Vol. 4, No. 7, 1986, pp. 1015–1020.
3. Heyel, C., The Encyclopaedia of Management, Van Nostrand Reinhold, New York, 1979.
4. Finch, W.L., Reliability: A Technical Management Challenge, Proceedings of the American Society for Quality Control Annual Conference, 1981, pp. 851–856.
5. Grant Ireson, W., Coombs, C.F., Moss, R.Y., Editors, Handbook of Reliability Engineering and Management, McGraw-Hill, New York, 1996.
6. Karger, D.W., Murdick, R.G., Managing Engineering Research, Industrial Press, New York, 1980.
7. Dhillon, B.S., Reiche, H., Reliability and Maintainability Management, Van Nostrand Reinhold, New York, 1985.
8. AMCP 706-104, Engineering Design Handbook: Value Engineering, Department of Defense, Washington, D.C., 1971.
9. Demarle, D.J., Shillito, M.L., Value Engineering, in Handbook of Industrial Engineering, edited by G. Salvendy, John Wiley and Sons, New York, 1982, pp. 7.3.1–7.3.20.
10. Younker, D.L., Value Engineering: Analysis and Methodology, Marcel Dekker, New York, 2003.
11. Hass, A.M.J., Configuration Management Principles and Practice, Addison-Wesley, Boston, 2003.
12. Riggs, J.L., Inoue, M.S., Introduction to Operations Research and Management Science: A General Systems Approach, McGraw-Hill, New York, 1975.

13. Antill, J.M., Critical Path Methods in Construction Practice, John Wiley and Sons, New York, 1982.
14. Thomas, E.F., Pitfalls in Reliability Program Management, Proceedings of the Annual Reliability and Maintainability Symposium, 1976, pp. 369–373.
15. Ekings, J.D., Ten Rules for the Reliability Professional, Proceedings of the American Society for Quality Control Annual Conference, 1982, pp. 343–351.

# chapter six

# Human and mechanical reliability

## 6.1 Introduction

Many times systems malfunction because of human errors rather than of software or hardware failures. The history of human reliability may be traced back to the late 1950s when H.L. Williams clearly pointed out that the reliability of the human-element must be included in the overall system reliability prediction; otherwise, such a prediction would not be realistic [1].

Over the years, many people have contributed to human reliability. The first book on the topic appeared in 1986 [2]. A comprehensive list of publications on human reliability is available in Ref. [3].

The history of mechanical reliability may be traced to World War II with the development of V1 and V2 rockets by the Germans. However, it was not until the mid-1960s when the mechanical reliability field received serious attention because of the loss of two spacecrafts (Syncom I and Mariner III) due to mechanical failures [4]. Consequently, the National Aeronautics and Space Administration (NASA) initiated many projects for improving reliability of mechanical items.

Nowadays, mechanical reliability has become an important element of the reliability field. A comprehensive list of publications on the topic is available in Ref. [5].

This chapter presents various important aspects of human reliability and mechanical reliability.

## 6.2 Human error occurrence facts and figures

Over the years, many studies concerning the occurrence of human errors have been performed. Results of some of these studies are as follows:

- A study of 23,000 defects in the production of nuclear parts revealed that about 82% of the defects were due to humans [6].
- Each year around 100,000 Americans die due to human errors in health care, and the annual financial impact on the U.S. economy is estimated to be somewhere between $17 billion and $29 billion [7].
- Over 90% of the documented air traffic control system errors were due to human operators [8].

- Over 50% of all technical medical equipment problems are caused by operator errors [9].
- A study of 135 vessel failures that took place during the period from 1926 to 1988 reported that around 25% of the failures were due to humans [10].
- A total of 401 human errors occurred in U.S. commercial light-water nuclear reactors during the period June 1, 1973 to June 30, 1975 [11].
- About 60% of all medical device-associated deaths and injuries reported through the Center for Devices and Radiological Health (CDRH) of the Food and Drug Administration (FDA) were caused by human errors [12].

## *6.3 Human error classifications and causes*

There are various types of human errors. They may be categorized under the following seven classifications [2, 13, 14]:

- **Classification I: Operator errors.** These errors are the result of operator mistakes and the conditions that lead to operator errors include complex tasks, poor environment, lack of proper procedures, operator carelessness, and poor personnel selection and training.
- **Classification II: Inspection errors.** These errors take place due to less than 100% accuracy of inspectors. One example of an inspection error is rejecting and accepting in-tolerance and out-of-tolerance items, respectively. Nonetheless, as per Ref. [15], an average inspection effectiveness is around 85%.
- **Classification III: Design errors.** These errors occur due to inadequate design. The causes of design errors include failure to implement human needs in the design, failure to ensure the man-machine interaction effectiveness, and assigning inappropriate functions to humans. An example of design errors is the placement of displays and controls so far apart that an operator is unable to use them effectively.
- **Classification IV: Assembly errors.** These errors take place during product assembly due to humans. These errors occur due to causes such as inadequate illumination, excessive temperature in the work area, poorly designed work layout, poor communication of related information, poor blueprints and other related material, and excessive noise level.
- **Classification V: Handling errors.** These errors take place due to inadequate transportation or storage facilities. More clearly, such facilities are not specified by the equipment manufacturer.
- **Classification VI: Installation errors.** These errors take place due to various reasons including using the incorrect installation associated instructions or blueprints or simply failing to install equipment according to the manufacturer's specification.

- **Classification VII: Maintenance errors.** These errors take place in the field due to maintenance personnel's oversights. As the equipment becomes old, the likelihood of the occurrence of these errors may increase due to the increase in maintenance frequency. Some examples of these errors are repairing the failed equipment incorrectly, applying the wrong grease at appropriate points of the equipment, and calibrating equipment incorrectly.

There are many causes for the occurrence of human errors. Some of these are as follows [2, 13]:

- Poor equipment design.
- Poor work layout.
- Poor motivation of involved personnel.
- Poor job environment: high noise level, poor lighting, crowded work space, high or low temperature, etc.
- Poor skill or training of all involved personnel.
- Poorly written or inadequate equipment operating and maintenance procedures.
- Complex tasks.
- Inadequate work tools.

## 6.4 Human stress-performance effectiveness and stress factors

Over the years, researchers have studied the relationship between stress and the effectiveness of human performance. The resulting curve of their effort is shown in Fig. 6.1 [13, 16]. This curve shows that a moderate level

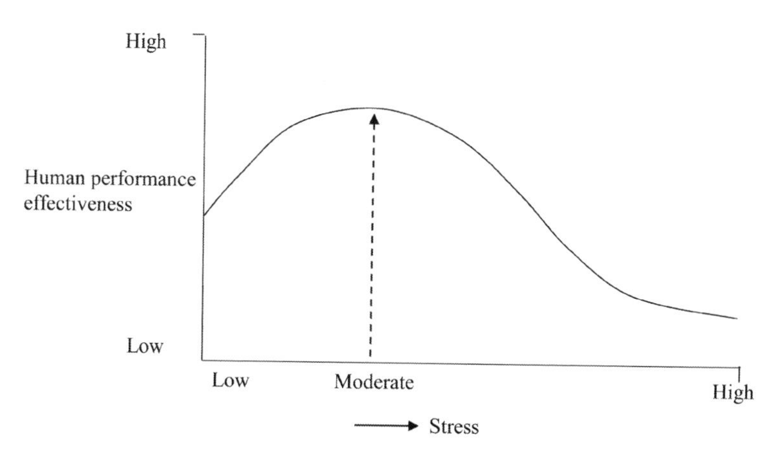


***Figure 6.1*** A hypothetical human stress-performance effectiveness curve.

of stress is required to increase the effectiveness of human performance to its maximum.

The moderate level may simply be interpreted as high enough stress for keeping the involved individual alert. At a very low stress, the task becomes unchallenging and dull; therefore, most humans will not conduct effectively, and their performance will not be at the optimum level. In contrast, as the stress bypasses its moderate level, the effectiveness of the human performance begins to decline. This decline is due to factors such as fear, worry, and other types of psychological stress. At the highest stress level, human reliability is at its lowest level.

There are many factors that can increase the stress on humans and in turn decrease their reliability in work and other environments. Some of these factors are as follows [16]:

- Dissatisfaction with the job.
- Inadequate expertise to perform the task.
- Performing tasks under extremely tight time schedules.
- Excessive demands of superiors.
- Working with people having unpredictable temperaments.
- Possibility of work layoff.
- Low chance of promotion from current position/job.
- Health-related problems.
- Serious financial problems.
- Difficulties with children/spouse.

## 6.5 *Human performance reliability in continuous time and mean time to human error (MTTHE) measure*

Humans perform various types of time-continuous tasks. Some examples of these tasks are aircraft maneuvering, missile countdown, and scope monitoring. The following equation, developed the same way as the general reliability function, can be used for calculating the human performance reliability in continuous time [2]:

$$R_h(t) = e^{-\int_0^t \lambda_{th}(t)dt} \tag{6.1}$$

where

$R_h(t)$ is the human performance reliability at time t.

$\lambda_{th}(t)$ is the time-dependent human error rate.

By integrating Equation (6.1) over the time interval $[0,\infty]$, we get the following general expression for MTTHE [2]:

$$MTTHE = \int_0^\infty R_h(t)\,dt = \int_0^\infty \exp\left[-\int_0^t \lambda_{th}(t)\,dt\right]dt \tag{6.2}$$

where
MTTHE is the mean time to human error.

It is to be noted that Equation (6.2) can be used for obtaining MTTHE when times to human error follow any probability distribution. The following example demonstrates the application of Equations (6.1) and (6.2).

**Example 6.1**

Assume that a person is performing a time-continuous task and his/her times to human error are exponentially distributed. The person's error rate is 0.004 errors per hour. Calculate the person's mean time to error and reliability for a 6-hour mission.

Thus, in this case we have

$$\lambda_{th}(t) = \lambda_{th} = 0.004 \text{ errors/hour}$$
$$t = 6 \text{ hours}$$

where
$\lambda_{th}$ is the person's constant error rate.

By inserting the specified values into Equation (6.1), we obtain

$$R_h(6) = e^{-\int_0^6 (0.004)dt} = e^{-(0.004)(6)} = 0.9762$$

Similarly, by using the given data values in Equation (6.2), we get

$$MTTHE = \int_0^\infty \exp\left[-\int_0^6 (0.004)\,dt\right]dt = \frac{1}{0.004} = 250 \text{ hours}$$

Thus, the person's mean time to error and reliability are 250 hours and 0.9762, respectively.

## 6.6 Human reliability analysis methods

There are many methods that can be used to perform various types of human reliability analysis [2]. Each has benefits and drawbacks. Two of these methods considered quite useful to perform human reliability analysis are presented below.

### 6.6.1 Markov method

This method is widely used in the area of reliability engineering, and it can also be used to perform time-continuous human reliability analysis. The method is described in Chapter 4 and is subject to the following assumptions [17]:

- All occurrences are independent of each other.
- The probability of a transition occurrence from one state to another in finite time $\Delta t$ is given by $\lambda_h \Delta t$. The parameter $\lambda_h$ in our case is the constant human error rate.
- The probability of more than one transitional occurrence in finite time $\Delta t$ is negligible $\left[\text{i.e.,} (\lambda_h \Delta t)(\lambda_h \Delta t) \rightarrow 0\right]$.

The following example demonstrates the application of this method in human reliability work:

**Example 6.2**

Assume that $\lambda_h$ is the constant error rate of a person performing a time-continuous task. The state space diagram of the person performing the task is shown in Fig. 6.2. Develop probability expressions for the person conducting the task successfully and unsuccessfully (i.e., the person commits an error) at time t with the aid of Markov method. Also, develop an expression for the MTTHE committed by the person in question.

In Fig. 6.2, numerals in boxes denote corresponding states (i.e., 0: person performing his/her task successfully [normally], 1: person committed an error).

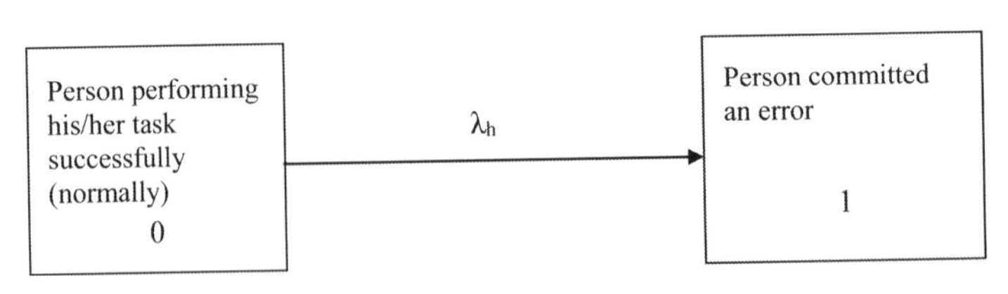


*Figure 6.2* State space diagram for a person conducting a time-continuous task.

The following symbols are associated with the Fig. 6.2 state space diagram:

- $\lambda_h$ is the constant error of the person (i.e., constant human error rate).
- $P_0(t)$ is the probability that the person is performing the task successfully (normally) at time t.
- $P_1(t)$ is the probability that the person has committed an error at time t.

With the aid of Markov method, we write down the following two Equations for the Fig. 6.2 state space diagram:

$$P_0(t+\Delta t) = P_0(t)(1-\lambda_h \Delta t) \tag{6.3}$$

$$P_1(t+\Delta t) = (\lambda_h \Delta t) P_0(t) + P_1(t) \tag{6.4}$$

where

$P_0(t+\Delta t)$ is the probability that the person is performing the task normally (successfully) at time $(t+\Delta t)$.

$P_1(t+\Delta t)$ is the probability that the person has committed an error at time t.

$(1-\lambda_h \Delta t)$ is the probability of the occurrence of no error in finite time interval $\Delta t$.

By rearranging Equations (6.3)–(6.4) and taking the limits, we obtain

$$\frac{dP_0(t)}{dt} + \lambda_h P_0(t) = 0 \tag{6.5}$$

$$\frac{dP_1(t)}{dt} - \lambda_h P_0(t) = 0 \tag{6.6}$$

At time t = 0, $P_0(0) = 1$ and $P_1(0) = 0$.

Solving Equations (6.5)–(6.6) using Laplace transforms [2] results in

$$P_0(t) = e^{-\lambda_h t} \tag{6.7}$$

$$P_1(t) = 1 - e^{-\lambda_h t} \tag{6.8}$$

Thus, the person's reliability is expressed by

$$R_p(t) = P_0(t) = e^{-\lambda_h t} \tag{6.9}$$

where

$R_p(t)$ is the person's reliability at time t.

By integrating Equation (6.9) over the time interval $[0, \infty]$, we obtain the following equation for the mean time to error committed by the person (MTTECP):

$$\begin{aligned} MTTECP &= \int_0^{\infty} R_p(t)\,dt \\ &= \int_0^{\infty} e^{-\lambda_h t}\,dt \\ &= \frac{1}{\lambda_h} \end{aligned} \tag{6.10}$$

Equations (6.7), (6.8), and (6.10) are the expressions for the probability of the person performing his/her task successfully, unsuccessfully, and the MTTECP, respectively.

**Example 6.3**

Assume that a person is performing a time-continuous task and his/her constant error rate is 0.002 errors per hour. Calculate the person's reliability during a 6-hour mission and mean time to error.

By substituting the given data into Equations (6.9) and (6.10), we obtain

$$\begin{aligned} R_p(6) &= e^{-(0.002)(6)} \\ &= 0.9880 \end{aligned}$$

and

$$\begin{aligned} MTTECP &= \frac{1}{0.002} \\ &= 500 \text{ hours} \end{aligned}$$

Thus, the person's reliability and mean time to error are 0.9880 and 500 hours, respectively.

## 6.6.2 *Fault tree analysis*

This method is often used to perform various types of reliability analysis, and it can also be used to perform human reliability analysis. The method is described in Chapter 4, and its application in human reliability work is demonstrated through the two examples presented below.

**Example 6.4**

Assume that a person is required to do a certain operation-related job: job M. The job is composed of two independent tasks N and X. If one of these two tasks is performed incorrectly, job M will not be accomplished successfully.

Task N is composed of three independent subtasks $n_1, n_2$, and $n_3$. For the successful performance of task N, only one of these three subtasks needs to be conducted correctly. Subtask $n_2$ is composed of two independent steps a and b. Both these steps must be performed correctly for the successful completion of subtask $n_2$.

Develop a fault tree for the top event: job M will not be accomplished correctly by the person.

Using the Chapter 4 fault tree symbols, the fault tree for the example is shown in Fig. 6.3.

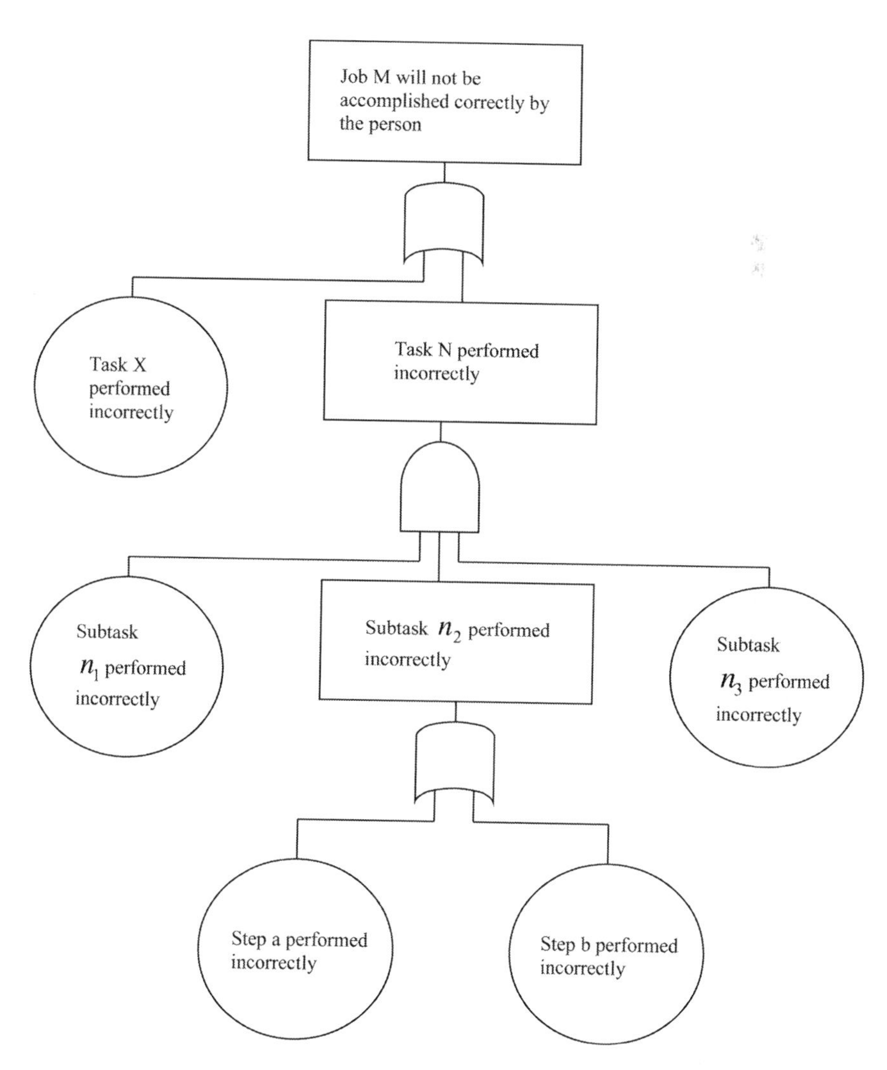


***Figure 6.3*** Fault tree for the top event: Job M will not be accomplished correctly.

**Example 6.5**

Assume that in Fig. 6.3, the occurrence probability of basic events $X, n_1, n_3, a,$ and $b$ is 0.04. Calculate the occurrence probability of the top event (i.e., job M will not be accomplished correctly) if all the events occur independently.

The probability of performing subtask $n_2$ incorrectly is given by [18]

$$P(n_2) = 1 - \{1 - P(a)\}\{1 - P(b)\}$$
$$= 1 - \{1 - 0.04\}\{1 - 0.04\}$$
$$= 0.0784$$

where
P(a) is the probability of performing step a incorrectly.
P(b) is the probability of performing step b incorrectly.

The probability of performing task N incorrectly is given by [18]

$$P(N) = P(n_1)P(n_2)P(n_3)$$
$$= (0.04)\,(0.0784)(0.04)$$
$$= 0.0001$$

The probability of accomplishing job M incorrectly is

$$P(M) = 1 - \{1 - P(x)\}\{1 - P(N)\}$$
$$= 1 - \{1 - 0.04\}\{1 - 0.0001\}$$
$$= 0.0401$$

Thus, the occurrence probability of the top event (i.e., job M will not be accomplished correctly) is 0.0401. Fig. 6.4 shows the Fig. 6.3 fault tree with calculated and given fault event occurrence probability values.

## 6.7 *Mechanical failure modes and general causes*

There are many different types of failure modes associated with mechanical parts/items. Some of these failure modes are shown in Fig. 6.5 [19–21]. These are bending failure, fatigue failure, bearing failure, metallurgical failure, instability failure, creep/rupture failure, compressive failure, shear loading failure, material flaw failure, stress concentration failure, ultimate tensile-strength failure, and tensile-yield-strength failure.

Bending failure occurs in situation when one outer surface is in compression and the other outer surface is in tension. An example of this

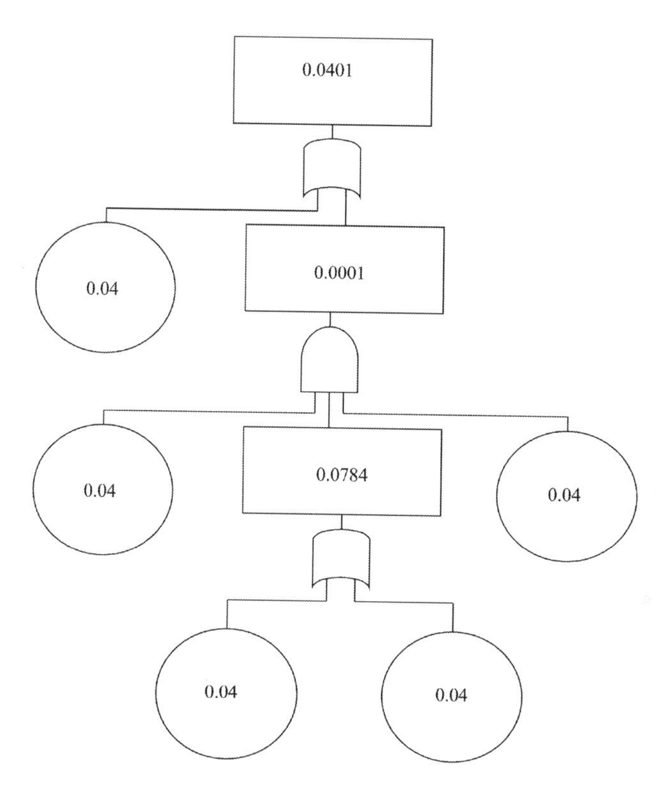


***Figure 6.4*** Fault tree with given and calculated fault event occurrence probability values.

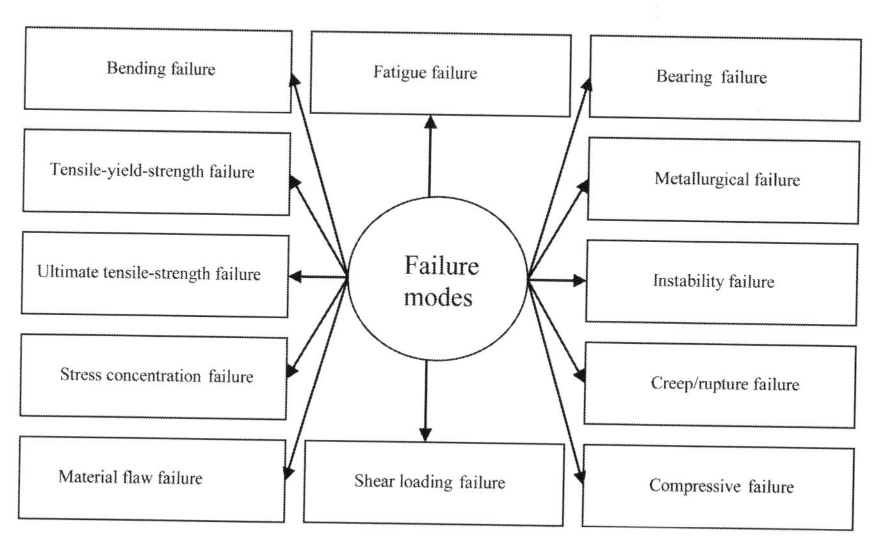


***Figure 6.5*** Mechanical failure modes.

failure is the tensile rupture of the outer material. In the case of fatigue failure, it occurs due to repeated loading or unloading (or partial unloading) of a part/item. The occurrence of this failure can be prevented by appropriate materials for a specific application. For example, steel outlasts aluminum under cyclic loading.

Bearing failure normally occurs due to a cylindrical surface bearing on either a flat or a concave surface like roller bearings in a race and is quite similar in nature to compressive failure. The metallurgical failure occurs because of extreme oxidation or operation in a corrosive environment. The occurrence of metallurgical failures is accelerated by environmental conditions such as erosion, corrosive media, heat, and nuclear radiation. It is to be noted that a metallurgical failure is also known as a material failure.

The instability failure is confined to structural members such as columns and beams, in particular, the ones manufactured using thin material where the loading is generally in compression. Nonetheless, this type of failure may also take place because of torsion or by combined loading (i.e., bending and compression). In the case of creep/rupture failure, material stretches (i.e., creeps) when the load is maintained on a continuous basis, and usually it ultimately terminates in a rupture. Furthermore, creep accelerates with elevated temperatures.

Compressive failure causes permanent deformation, cracking, or rupturing and is quite similar to tensile failure except under compressive loads. Shear loading failure occurs in situation when shear stress becomes higher than the material's strength when applying high shear or torsion loads.

Material flaw failure occurs due to factors such as small cracks and flaws, weld defects, fatigue cracks, and poor quality assurance. The stress concentration failure takes place under the conditions of uneven stress flow through a mechanical design.

Ultimate tensile-strength failure occurs in situations when the ultimate tensile strength is less than the applied stress and leads to a complete failure of the structure at a cross-sectional point. Finally, the tensile-yield-strength failure occurs under tension and, more clearly, when the applied stress is higher than the material yield strength.

Over the years, various people have studied the mechanical failures' causes and have identified them as follows [19]:

- Defect or poor design.
- Manufacturing defect.
- Incorrect installation.
- Wrong application.
- Gradual/deterioration in performance.
- Wear-out.
- Failure of other components/parts.

## 6.8 Safety factors and safety margin

Safety factors and safety margin are used during the design for ensuring reliability of mechanical items and, basically, are arbitrary multipliers. They can be useful for providing satisfactory design if they are established with care, using considerable past experiences and data.

Both safety factors and safety margin are presented next.

### 6.8.1 Safety factor I

This is defined by [20]

$$SF = \frac{FGST_m}{MFGS} \geq 1 \tag{6.11}$$

where

SF is the safety factor.
$FGST_m$ is the mean failure-governing strength.
MFGS is the mean failure-governing stress.

This safety factor is a quite good measure when both strength and stress are normally distributed. However, for large variations in stress and strength, it becomes meaningless because of the positive failure rate.

### 6.8.2 Safety factor II

This is defined by [21]

$$SF = \frac{MSL}{NSL} \tag{6.12}$$

where

SF is the safety factor.
MSL is the maximum safe load.
NSL is the normal service load.

This safety factor is considered quite useful, particularly when the loads are distributed.

### 6.8.3 Safety factor III

This is defined by [22]

$$SF = \frac{SM}{WS_{ma}} \tag{6.13}$$

where

SF is the safety factor.
SM is the strength of the material.
$WS_{ma}$ is the maximum allowable working stress.

The value of this safety factor is always greater than one. More clearly, when its value is less than one, it simply means that the item under consideration will malfunction because its strength is less than the applied stress.

The standard deviation of the safety factor is defined by [22].

$$\sigma = \left[ \left( \sigma_1 / WS_{ma} \right)^2 + \left\{ SM / \left( WS_{ma} \right)^2 \right\} \sigma_2^2 \right]^{1/2} \tag{6.14}$$

where

$\sigma$ is the standard deviation of the safety factor.
$\sigma_1$ is the standard deviation of the strength.
$\sigma_2$ is the standard deviation of the stress.

### 6.8.4 *Safety margin*

This is defined by [23]

$$\beta = SF - 1 \tag{6.15}$$

where

$\beta$ is the safety margin.
SF is the safety factor.

It is to be noted that the value of this measure is always greater than zero, and its negative value indicates that the item under consideration will malfunction.

For normally distributed strength and stress, the safety margin is defined by [23]

$$\beta = \left( \theta_a - \theta_{ms} \right) / \sigma_{sth} \tag{6.16}$$

where

$\theta_a$ is the average strength.
$\theta_{ms}$ is the maximum stress.
$\sigma_{sth}$ is the standard deviation of the strength.

The maximum stress is expressed by

$$\theta_{ms} = \theta_m + k\sigma_{st} \tag{6.17}$$

where

$\theta_m$ is the mean value of the stress.
k is a factor that takes values between 3 and 6.
$\sigma_{st}$ is the standard deviation of the stress.

## 6.9 *Stress–strength interference theory modeling*

When the probability density function of stress and strength of an item are known, its reliability can be determined analytically. The method used to determine reliability is referred to as stress–strength interference theory modeling. In this case, reliability is simply the probability that the failure-governing stress will not exceed the failure-governing strength. Mathematically, reliability is expressed by [24, 25]

$$R = P(y > x) = P(x < y) \tag{6.18}$$

where

R is the item reliability.
P is the probability.
y is the strength random variable.
x is the stress random variable.

Using Ref. [25], we rewrite Equation (6.18) as follows:

$$R = \int_{-\infty}^{\infty} f(x)\left[\int_{x}^{\infty} f(y)dy\right]dx \tag{6.19}$$

where

f(y) is the probability density function of the strength.
f(x) is the probability density function of the stress.

It is to be noted that in the published literature, Equation (6.19) is also written in the following three forms [25]:

$$R = \int_{-\infty}^{\infty} f(x)\left[1 - \int_{-\infty}^{x} f(y)dy\right]dx \tag{6.20}$$

$$R = \int_{-\infty}^{\infty} f(y)\left[\int_{-\infty}^{y} f(x)dx\right]dy \tag{6.21}$$

$$R = \int_{-\infty}^{\infty} f(y)\left[1 - \int_{y}^{\infty} f(x)dx\right]dy \tag{6.22}$$

The application of Equation (6.19) is demonstrated by developing the following stress–strength model when the strength and stress probability density functions of an item are known:

### 6.9.1 Model

In this case, strength and stress associated with an item are exponentially distributed. Thus, we have

$$f(y) = \alpha e^{-\alpha y}, 0 \le y \le \infty \tag{6.23}$$

and

$$f(x) = \mu e^{-\mu x}, 0 \le x \le \infty \tag{6.24}$$

where

$\alpha$ and $\mu$ are the reciprocals of the mean values of strength and stress, respectively.

By inserting Equations (6.23) and (6.24) into Equation (6.19), we obtain

$$\begin{aligned} R &= \int_0^\infty \mu e^{-\mu x} \left[ \int_x^\infty \alpha e^{-\alpha y} dy \right] dx \\ &= \int_0^\infty \mu e^{-(\mu+\alpha)x} dx \\ &= \frac{\mu}{\mu + \alpha} \end{aligned} \tag{6.25}$$

For $\alpha = \frac{1}{\bar{y}}$ and $\mu = \frac{1}{\bar{x}}$, Equation (6.25) becomes

$$R = \frac{\bar{y}}{\bar{y} + \bar{x}} \tag{6.26}$$

where

$\bar{y}$ and $\bar{x}$ are the mean strength and stress, respectively.

**Example 6.6**

Assume that strength and stress of an item are exponentially distributed with mean values of 25,000 psi and 5,000 psi, respectively. Calculate the item reliability.

By inserting the given data values into Equation (6.26), we obtain

$$R = \frac{25,000}{25,000 + 5,000}$$

$$= 0.8333$$

Thus, the item reliability is 0.8333.

## 6.10 Failure rate models

Over the years, many mathematical models have been developed to estimate failure rates of mechanical items such as pumps, brakes, filters, bearings, and compressors [26–29]. This section presents three of these failure rate estimation models.

### 6.10.1 Pump failure rate model

The pump failure rate is defined by [26]

$$\lambda_p = \lambda_{fd} + \lambda_s + \lambda_b + \lambda_{se} + \lambda_c \tag{6.27}$$

where

$\lambda_p$ is the pump failure rate, expressed in failures/$10^6$cycles.
$\lambda_{fd}$ is the pump fluid driver failure rate.
$\lambda_s$ is the pump shaft failure rate.
$\lambda_b$ is the pump bearings' failure rate.
$\lambda_{se}$ is the pump seals' failure rate.
$\lambda_c$ is the pump casing failure rate.

The pump shaft failure rate, $\lambda_s$, is expressed by

$$\lambda_s = \lambda_{sb} \prod_{j=1}^{6} f_j \tag{6.28}$$

where

$\lambda_{sb}$ is pump shaft base failure rate.
$f_j$ is the jth modifying factor; for j = 1 is associated with material temperature, j = 2 pump displacement, j = 3 casing thrust load, j = 4 shaft surface finish, j = 5 contamination, and j = 6 material endurance limit.

The pump seal failure rate, $\lambda_{se}$, is defined by

$$\lambda_{se} = \lambda_{seb} \prod_{j=1}^{7} f_j \tag{6.29}$$

where

$\lambda_{seb}$ is the pump seal base failure rate.

$f_j$ is the jth modifying factor; for j = 1 is for effects of casing thrust load, j = 2 surface finish, j = 3 seal smoothness, j = 4 fluid viscosity, j = 5 pressure/velocity factor, j = 6 temperature, and j = 7 contaminates.

Similarly, the values of $\lambda_{fd}$, $\lambda_b$, and $\lambda_c$ can be calculated. Procedures for calculating these failure rates are given in Ref. [26].

### 6.10.2 Brake system failure rate model

The brake system failure rate is defined by [28]

$$\lambda_{bs} = \lambda_{bh} + \lambda_{bf} + \lambda_{sp} + \lambda_{ss} + \lambda_{be} + \lambda_{as} \tag{6.30}$$

where

$\lambda_{bs}$ is the brake system failure rate, expressed in failures/$10^6$hours.

$\lambda_{bh}$ is the brake housing failure rate.

$\lambda_{bf}$ is the brake friction materials' failure rate.

$\lambda_{sp}$ is the springs' failure rate.

$\lambda_{ss}$ is the seals' failure rate.

$\lambda_{be}$ is the bearings' failure rate.

$\lambda_{as}$ is the actuators' failure rate.

The values $\lambda_{bh}$, $\lambda_{bf}$, $\lambda_{sp}$, $\lambda_{ss}$, $\lambda_{be}$, and $\lambda_{as}$ are estimated through various means [28, 30]. Additional information on this model is available in Ref. [18].

### 6.10.3 Filter failure rate model

The following equation can be used for predicting filter failure rate [26]:

$$\lambda_{ff} = \lambda_{bff} \prod_{j=1}^{6} f_j \tag{6.31}$$

where

$\lambda_{ff}$ is the filter failure rate, expressed in failures/$10^6$hours.

$\lambda_{bff}$ is the filter base failure rate.

$f_j$ is the jth modifying factor; for j = 1 is for temperature effects, j = 2 vibration effects, j = 3 water contamination effects, j = 4 cyclic flow effects, j = 5 cold start effects, and j = 6 differential pressure effects.

Additional information on this model is available in Ref. [26].

## 6.11 Problems

1. Discuss the following classifications of human errors:
    - Assembly errors.
    - Operator errors.
    - Design errors.
2. What are the causes for the occurrence of human errors?
3. Describe the human performance effectiveness versus stress curve.
4. What are the factors that can increase stress on humans?
5. Assume that a person is performing a time-continuous task and his/her times to human error are exponentially distributed. The person's error rate is 0.005 errors per hour. Calculate the person's MTTHE and reliability for an 8-hour mission.
6. List at least 12 mechanical failure modes.
7. What are the general causes for the occurrence of mechanical failures?
8. Define the safety factor that is considered a good measure when both stress and strength are normally distributed.
9. Define the safety margin when both stress and strength are normally distributed.
10. Assume that stress and strength of an item are exponentially distributed with mean values of 4000 psi and 30,000 psi, respectively. Calculate the item reliability.

## References


1. Williams, H.L., Reliability Evaluation of the Human Component in Man-Machine Systems, Electrical Manufacturing, April 1958, pp. 78–82.
2. Dhillon, B.S., Human Reliability: With Human Factors, Pergamon Press, New York, 1986.
3. Dhillon, B.S., Yang, N., Human Reliability: A Literature Survey and Review, Microelectronics and Reliability, Vol. 34, 1994, pp. 803–810.
4. Redler, W.M., Mechanical Reliability Research in the National Aeronautics and Space Administration, Proceedings of the Reliability and Maintainability Conference, 1966, pp. 763–768.
5. Dhillon, B.S., Reliability and Quality Control: Bibliography on General and Specialized Areas, Beta Publishers, Gloucester, Ontario, Canada, 1992.
6. Rook, L.W., Reduction of Human Error in Industrial Production, Report No. SCTM 93-63(14), Sandia National Laboratories, Albuquerque, New Mexico, 1962.
7. Kohn, L.T., Corrigan, J.M., Donaldson, M.S., To Err is Human: Building a Safer Health System, Institute of Medicine Report, National Academy Press, Washington, D.C., 1999.

8. Kenney, G.C., Span, M.J., Amato, R.A., The Human Element in Air Traffic Control: Observations and Analysis of Controllers and Supervisors in Providing Air Traffic Control Separation Services, Report No. MTR-7655, METREK Div., MITRE Corporation, 1977.
9. Dhillon, B.S., Reliability Technology in Health Care Systems, Proceedings of the IASTED International Symposium on Computers and Advanced Technology in Medicine, Health Care, Bioengineering, 1990, pp. 84–87.
10. Organizational Management and Human Factors in Quantitative Risk Assessment, Report No. 33/1992 (Report 1), British Health and Safety Executive (HSE), London, 1992.
11. Joos, D.W., Sabri, Z.A., Husseiny, A.A., Analysis of Gross Error Rates in Operation of Commercial Nuclear Power Stations, Nuclear Engineering Design, Vol. 52, 1979, pp. 265–300.
12. Bogner, M.S., Medical Devices: A New Frontier for Human Factors, CSERIAC Gateway, Vol. 4, No. 1, 1993, pp. 12–14.
13. Meister, D., The Problem of Human-initiated Failures, Proceedings of the Eight National Symposium on Reliability and Quality Control, 1962, pp. 234–239.
14. Cooper, J.I., Human-Initiated Failures and Man-Function Reporting, IRE Trans. Human Factors, Vol. 10, 1961, pp. 104–109.
15. McCornack, R.L., Inspector Accuracy: A Study of the Literature, Report No. SCTM 53–61 (14), 1961, Sandia Corporation, Albuquerque, New Mexico.
16. Beech, H.R., Burns, L.E., Sheffield, B.F., A Behavioral Approach to the Management of Stress, John Wiley and Sons, New York, 1982.
17. Shooman, M.L., Probabilistic Reliability: An Engineering Approach, McGraw-Hill, New York, 1968.
18. Dhillon, B.S., Design Reliability: Fundamentals and Applications, CRC Press, Boca Raton, Florida, 1999.
19. Lipson, C., Analysis and Prevention of Mechanical Failures, Course Notes No. 8007, University of Michigan, Ann Arbor, Michigan, 1980.
20. Bompass-Smith, J.H., Mechanical Survival: The Use of Reliability Data, McGraw-Hill, London, 1973.
21. Phelan, M.R., Fundamentals of Machine Design, McGraw-Hill, New York, 1962.
22. Grant Irson, W., Coombs, C.F., Moss, R.Y., Handbook of Reliability Engineering and Management, McGraw-Hill, New York, 1996.
23. Keceioglu, D., Haugen, E.B., A Unified Look at Design Safety Factors, Safety Margins, and Measures of Reliability, Proceedings of the Annual Reliability Maintainability Conference, 1968, pp. 522–530.
24. Keceioglu, D., Reliability Analysis of Mechanical Components and Systems, Nuclear Engineering and Design, Vol. 19, 1972, pp. 249–290.
25. Dhillon, B.S., Mechanical Reliability: Theory, Models, and Applications, American Institute of Aeronautics and Astronautics, Washington, D.C., 1988.
26. Raze, J.D., Nelson, J.J., Simard, D.J., Bradley, M., Reliability Models for Mechanical Equipment, Proceedings of the Annual Reliability and Maintainability Symposium, 1987, pp. 130–134.
27. Grant Ireson, W., Coombs, C.F., Moss, R.Y., Handbook of Reliability Engineering and Management, McGraw-Hill, New York, 1996.

28. Rhodes, S., Nelson, J.J., Raze, J.D., Bradley, M., Reliability Models for Mechanical Equipment, Proceedings of the Annual Reliability and Maintainability Symposium, 1988, pp. 127–131.
29. Nelson, J.J., Raze, J.D., Bowman, J., Perkins, G., Wannamaker, A., Reliability Models for Mechanical Equipment, Proceedings of the Annual Reliability and Maintainability Symposium, 1989, pp. 146–153.
30. Boone, T.D., Reliability Prediction Analysis for Mechanical Brake Systems, NAVAIR/SYSCOM Report, August 1981, Department of Navy, Department of Defense, Washington, D.C.

# *chapter seven*

# *Reliability testing and growth*

## *7.1 Introduction*

Reliability testing is one of the most important reliability activities of a reliability program. Its main objective is to obtain information concerning failures, in particular, the equipment/product tendency to fail as well as failure consequences. This type of information is very useful to control failure tendencies along with their consequences. A good reliability test program may simply be described as the one that provides maximum information concerning failures from minimal amount of testing. Over the years, many publications on reliability testing have appeared; in particular, two such publications are listed in Refs. [1, 2].

In the design and development of new systems, the first prototypes generally contain various design and engineering-related deficiencies. Elimination of such deficiencies leads to a system's/product's reliability growth [3]. Although the serious thinking concerning reliability growth may be traced back to the late 1950s, the frequently used reliability growth monitoring model was postulated by Duane in 1964 [4]. A comprehensive list of publications up to 1980 on reliability growth is available in Ref. [5].

This chapter presents various important aspects of reliability testing and growth.

## *7.2 Reliability test classifications*

Reliability tests may be grouped under the following three classifications [6, 7]:

- **Reliability development and demonstration testing.** This is concerned with meeting objectives such as to indicate whether any design changes are needed, to determine whether the design is to be improved to meet the stated reliability requirement, as well as to verify improvements in design reliability. This type of testing depends on factors such as the type of system/subsystem being investigated and the level of complexity under consideration. For example, in the case of electronic parts, the reliability development and demonstration testing could take the form of life tests to evaluate whether the part can meet its reliability-related goals; if not, what actions are needed?

In order to meet objectives of reliability development and demonstration testing effectively, the accumulated test data must be a kind that clearly allow insight into the failure effects/probabilities for a certain design under consideration. Furthermore, these data values serve as a good basis for reliability assessment and analysis for two specific items: design under consideration and subsequent related programs.

- **Qualification and acceptance testing.** This is concerned with meeting two fundamental objectives: to determine whether a certain design is qualified for its projected application and to arrive at a decision whether a part/assembly/end item is to be accepted or not. These objectives differ from the objectives of other reliability tests, particularly with respect to reject/accept mechanism. Furthermore, qualification and acceptance testing incorporates the usual testing and screening the quality-control function of incoming parts. In regard to the materials and components to be used in the equipment/system under development, the qualification and acceptance testing begins early in the program.
- **Operational testing.** This is concerned with objectives that include verifying the results of reliability analysis performed during the equipment/system design and development, providing data for subsequent activities and providing data indicating desirable changes to operating policies and procedures in regard to reliability/maintainability. Finally, it is added that the operational testing provides the feedback from practice to theory.

## 7.3 Success testing

This type of testing is sometimes used in receiving inspection and in engineering test laboratories where a no-failure test is stated. Normally, the main goal for this test is to ensure that a certain reliability level has been achieved at a stated confidence level.

In this case, for zero failures, the lower $100(1-\theta)\%$ confidence limit on the desired reliability level is expressed as [8]

$$R_d = \theta^{1/n} \tag{7.1}$$

where

θ is the level of significance or consumer's risk.

n is the number of items placed on test.

Thus, with $100(1-\theta)\%$ confidence, it may be stated that

$$R_d \le R_t \tag{7.2}$$

where

$R_t$ is the true reliability.

By taking the natural logarithms of the both sides of Equation (7.1), we obtain

$$\ln R_d = \frac{1}{n} \ln \theta \tag{7.3}$$

Thus, from Equation (7.3), we get

$$n = \frac{\ln \theta}{\ln R_d} \tag{7.4}$$

The desired confidence level, $CL_d$, is expressed by

$$CL_d = 1 - \theta \tag{7.5}$$

By rearranging Equation (7.5), we obtain

$$\theta = 1 - CL_d \tag{7.6}$$

Using Equations (7.2) and (7.6) in Equation (7.4), we obtain

$$n = \frac{\ln(1 - CL_d)}{\ln R_t} \tag{7.7}$$

Equation (7.7) can be used for determining the number of items to be tested for stated reliability and confidence level.

**Example 7.1**

Assume that 90% reliability of an electronic item is to be demonstrated at 95% confidence level. Calculate the number of electronic items to be placed on test when no failures are allowed.

By substituting the specified data values into Equation (7.7), we get

$$n = \frac{\ln(1 - 0.95)}{\ln 0.9}$$

$$= 28$$

Thus, 28 electronic items must be placed on test.

## 7.4 Accelerated life testing

This is used for obtaining quick information on item's life distributions, reliabilities, failure rates, etc., by subjecting the test items to conditions such that the failures occur earlier. Thus, the accelerated life testing is a quite useful tool for making long-term reliability prediction within a short time span. The following two methods are used for performing an accelerated life test [7, 9–11]:

- **Method I.** This is concerned with carrying out the test at very high stress (e.g., voltage, temperature, and humidity) levels so that malfunctions can be induced in a very short time interval. Typical examples of items for application under this method are communication satellites, air traffic control monitors, and components of a power-generating unit. Generally, accelerated failure time testing over the accelerated stress testing is preferred because there is no need for making assumptions concerning the relationship of time to failure distributions at both normal and accelerated conditions. Nonetheless, the results of the accelerated-stress testing are related to the normal conditions by using various mathematical models. One such model will be presented subsequently in this section.
- **Method II.** This method is concerned with accelerating the test by using the item under consideration more intensively than its general usage. Normally, the items such as a high bulb of a telephone set and a crank shaft of a car used discretely or non-discretely can be tested by employing this method. However, it is to be noted that it is not possible to use this method for an item such as a mainframe computer in constant use. Under such a scenario, method I can be used.

### 7.4.1 Relationship between the accelerated and normal conditions

This section presents relationships between the accelerated and normal conditions for the following four items [7, 12]:

- **Probability density function**

  The normal operating condition failure probability density function is defined by

$$f_n(t) = \frac{1}{\alpha} f_{st}\left(\frac{t}{\alpha}\right) \tag{7.8}$$

where

$\alpha$ is the acceleration factor.

t is time.

$f_n(t)$ is the normal operating condition failure probability density function.

$f_{st}\left(\frac{t}{\alpha}\right)$ is the stressful operating condition failure probability density function.

- **Cumulative distribution function**

  The normal operating condition cumulative distribution function is expressed by

$$F_n(t) = F_{st}\left(\frac{t}{\alpha}\right) \tag{7.9}$$

where

$F_{st}\left(\frac{t}{\alpha}\right)$ is the stressful operating condition cumulative distribution function.

- **Hazard rate**

  The normal operating condition hazard rate is expressed by

$$h_n(t) = \frac{f_n(t)}{1 - F_n(t)} \tag{7.10}$$

By substituting Equations (7.8) and (7.9) into Equation (7.10), we get

$$h_n(t) = \frac{\frac{1}{\alpha} f_{st}\left(\frac{t}{\alpha}\right)}{1 - F_{st}\left(\frac{t}{\alpha}\right)} \tag{7.11}$$

Thus, from Equation (7.11), we have

$$h_n(t) = \frac{1}{\alpha} h_{st}\left(\frac{t}{\alpha}\right) \tag{7.12}$$

where

$h_{st}\left(\frac{t}{\alpha}\right)$ is the stressful operating condition hazard rate.

- **Time to failure**

  The time to failure at normal operating condition is expressed by

$$t_n = \alpha t_{st} \tag{7.13}$$

where

$t_n$ is the time to failure at normal operating condition.
$t_{st}$ is the time to failure at stressful operating condition.

### 7.4.2 Acceleration model

For an exponentially distributed time to failure at an accelerated stress, s, the cumulative distribution function is expressed by

$$F_{st}(t) = 1 - e^{-\lambda_{st} t} \tag{7.14}$$

where
$\lambda_{st}$ is the constant failure rate at the stressful level.

Thus, from Equations (7.9) and (7.14), we obtain

$$F_n(t) = F_{st}\left(\frac{t}{\alpha}\right) = 1 - e^{-\lambda_{st} t/\alpha} \tag{7.15}$$

Similarly, using Equation (7.12), we have

$$\lambda_n = \frac{\lambda_{st}}{\alpha} \tag{7.16}$$

where
$\lambda_n$ is the constant failure rate at the normal operating conditions.

For both non-censored and censored data, the failure rate at the stressful level can be calculated by using the following two equations, respectively [7, 13]:

- **Non-censored data**

$$\lambda_{st} = m \Big/ \sum_{i=1}^{m} t_i \tag{7.17}$$

where
$t_i$ is the ith failure time; i = 1,2,3,…,m.
m is the total number of items under test at a certain stress.

- **Censored data**

$$\lambda_{st} = n \Big/ \left( \sum_{i=1}^{n} t_i + \sum_{i=1}^{m-n} t_i^{/} \right) \tag{7.18}$$

where
n is the number of failed items at the accelerated stress.
$t_i^{/}$ is the ith censoring time.

**Example 7.2**

Assume that a sample of 30 identical electronic items was accelerated life tested at 120°C, and their times to failure were exponentially distributed with a mean value of 6000 hours. If the value of the acceleration factor is 20 and the electronic items' normal operating temperature is 30°C, calculate the electronic items, operating at the normal conditions, failure rate, mean time to failure, and reliability for a 4000-hour mission.

In this case, the electronic items' failure rate at the accelerated temperature is expressed by

$$\lambda_{st} = 1/\left(\text{electronic items' mean life under accelerated testing}\right)$$
$$= \frac{1}{6000} = 0.000166 \text{ failures/hour}$$

By substituting the above result and the given data into Equation (7.16), we obtain

$$\lambda_n = \frac{0.000166}{20}$$
$$= 8.3333 \text{ failures/hour}$$

Thus, the electronic items' mean time to failure $(MTTF_n)$ at the normal operating condition is

$$MTTF_n = \frac{1}{\lambda_n}$$
$$= 120{,}000 \text{ hours}$$

The electronic items' reliability for a 4000-hour mission at the normal operating condition is

$$R(4{,}000) = e^{-\left(8.3333\times10^{-6}\right)(4{,}000)}$$
$$= 0.9672$$

Thus, the electronic items' failure rate, mean time to failure, and reliability at the normal operation are $8.3333 \times 10^{-6}$ failure/hour, 120,000 hours, and 0.9672, respectively.

## 7.5 *Confidence interval estimates for mean time between failures*

Usually, in reliability studies in the industrial sector, the time to item failure is assumed to be exponentially distributed. Thus, the item failure rate becomes constant and, in turn, the mean time between failures (MTBF) is simply the reciprocal of the failure rate.

In testing a sample of parts/items with exponentially distributed times to failures, a point estimate of MTBF can be made; however, this figure only provides an incomplete picture because it fails to provide any surety of measurement. Nonetheless, it would probably be more realistic if we say, for example, that after testing a sample of items for t hours, n number of failures have occurred and the MTBF lies somewhere between certain lower and upper limits with certain confidence.

The confidence intervals on MTBF can be calculated by utilizing the $\chi^2$ (chi-square) distribution. The usual notation used for obtaining chi-square values is as follows:

$$\chi^2(p, df) \tag{7.19}$$

where

df is the degrees of freedom.
p is a quantity function of the confidence coefficient.

The following list of symbols is used in subsequent associated formulas in this section [6, 7]:

$\theta$ is the acceptable error risk.
$\beta$ is the mean life or MTBF.
$C = 1 - \theta$ is the confidence level.
n is the number of items that were placed on test at zero time (i.e., $t = 0$).
m is the number of failures accumulated to time $t^*$, where $t^*$ denotes the life test termination time.
$m^*$ is the number of preassigned failures.

There are following two different cases to estimate confidence intervals:

- Testing is terminated at a preassigned time $t^*$.
- Testing is terminated at a preassigned number of failures, $m^*$.

Thus, for the above two cases, to compute upper and lower limits of MTBF, the following formulas can be used [6, 7]:

- Preassigned truncation time, $t^*$

$$\left[\frac{2Y}{\chi^2\left(\frac{\theta}{2}, 2m+2\right)}, \frac{2Y}{\chi^2\left(1-\frac{\theta}{2}, 2m\right)}\right] \tag{7.20}$$

- Preassigned number of failures, $m^*$

$$\left[\frac{2Y}{\chi^2\left(\frac{\theta}{2},2m\right)},\frac{2Y}{\chi^2\left(1-\frac{\theta}{2},2m\right)}\right] \quad (7.21)$$

It is to be noted that the value of Y is determined by the test types: replacement test (i.e., the failed unit is replaced or repaired) and non-replacement test.

Thus, for the replacement test, we have

$$Y = nt^* \quad (7.22)$$

Similarly, for the non-replacement test, we have

$$Y = (n-m)t^* + \sum_{j=1}^{m} t_j \quad (7.23)$$

where

$t_j$ is the jth failure time.

In the case of censored items/units (i.e., withdrawal or loss of unfailed items/units), the value of Y becomes as follows:

- For replaced failed units/items but non-replacement of censored units/items

$$Y = (n-c)t^* + \sum_{i=1}^{c} t_i \quad (7.24)$$

where

c is the number of censored items/units.

$t_i$ is the ith, censorship time.

- For non-replaced failed and censored items/units

$$Y = (n-c-m)t^* + \sum_{i=1}^{c} t_i + \sum_{j=1}^{m} t_j \quad (7.25)$$

Some tabulated values of $\chi^2(p,df)$ are presented in Table 7.1.

***Table 7.1*** Values of chi-square distribution

| Degrees of freedom | $\chi^2$ value | | | |
|---|---|---|---|---|
| 2 | 0.1 | 0.21 | 4.6 | 5.99 |
| 4 | 0.71 | 1.06 | 7.77 | 9.44 |
| 6 | 1.63 | 2.2 | 10.64 | 12.59 |
| 8 | 2.73 | 3.49 | 13.36 | 15.5 |
| 10 | 3.94 | 4.86 | 15.98 | 18.3 |
| 12 | 5.22 | 6.3 | 18.54 | 21.02 |
| 14 | 6.57 | 7.79 | 21.06 | 23.68 |
| 16 | 7.96 | 9.31 | 23.54 | 26.29 |
| 18 | 9.39 | 10.86 | 25.98 | 28.86 |
| 20 | 10.85 | 12.44 | 28.41 | 31.41 |
| Probability value: | 0.95 | 0.9 | 0.1 | 0.05 |

**Example 7.3**

Assume that 20 identical electronic items were placed on test at time $t = 0$ and none of the failed items were replaced and the test was terminated after 200 hours. Three electronic items failed after 20, 40, and 60 hours of operation. Calculate the electronic items' MTBF and its upper and lower limits with 90% confidence level.

By substituting the given data values into Equation (7.23), we get

$$Y = (20-3)(200) + (20+40+60) = 3520 \text{ hours}$$

Thus, the electronic items' MTBF is given by

$$\beta = \frac{3520}{3} = 1173.3 \text{ hours}$$

By substituting the given and other values into Equation (7.20) and using Table 7.1, we obtain the following values of MTBF upper and lower limits:

$$\text{Upper limit} = \frac{2(3520)}{\chi^2(0.95,6)}$$

$$= \frac{2(3520)}{1.63}$$

$$= 4319.01 \text{ hours}$$

$$\text{Lower limit} = \frac{2(3520)}{\chi^2(0.05,8)}$$

$$= \frac{2(3520)}{15.5}$$

$$= 454.19 \text{ hours}$$

Thus, we can state with 90% confidence that the electronic items' true MTBF will lie within 454.19 hours and 4319.01 hours or $454.19 \leq \beta \leq 4319.01$.

**Example 7.4**

Assume that 16 electronic items were put on test at zero time and at the occurrence of eighth failure, the testing was stopped. The eighth failure occurred at 100 hours and all the failed items were replaced.

Calculate the MTBF of the electronic items and upper and lower limits on MTBF at 80% confidence level.

By substituting the given data values into Equation (7.22), we obtain

$$Y = (16)(100) = 1600 \text{ hours}$$

Thus, the electronic items' MTBF is given by

$$\beta = \frac{1600}{8} = 200 \text{ hours}$$

By substituting the given and other values into Equation (17.21) and using Table 7.1, we obtain the following values of MTBF upper and lower limits:

$$\text{Upper limit} = \frac{2(1600)}{\chi^2(0.90, 16)}$$

$$= \frac{2(1600)}{9.31}$$

$$= 343.7 \text{ hours}$$

$$\text{Lower limit} = \frac{2(1600)}{\chi^2(0.1, 16)}$$

$$= \frac{3200}{23.54}$$

$$= 135.9 \text{ hours}$$

Thus, at 80% confidence level, the electronic items' true MTBF will lie within 135.9 hours and 343.7 hours or $135.9 \leq \beta \leq 343.7$.

## *7.6 Reliability growth program and reliability growth process evaluation approaches*

The reliability growth program is a structured process used for discovering reliability deficiencies through testing, analyzing such deficiencies, and implementation of corrective measures for lowering the rate

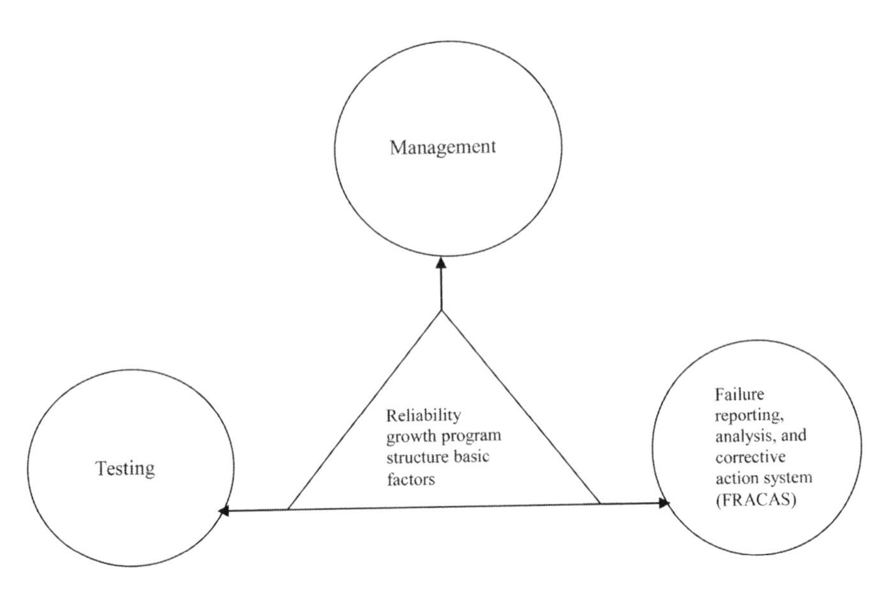


*Figure 7.1* Reliability growth program structure basic factors.

of occurrence. Some of the important benefits of the reliability growth program include assessments of achievement and projecting the product reliability trends. Nonetheless, the fundamental structure of a reliability growth program is composed of three basic factors, as shown in Fig. 7.1

These factors are testing; failure reporting, analysis, and corrective action system (FRACAS); and management. Testing provides opportunity for discovering deficiencies in areas such as design, manufacturing, procedures, and components. FRACAS is basically a process employed for determining failure root causes and recommending necessary corrective actions. Finally, management is concerned with making it happen, and some of the major management considerations concerning a reliability growth program are as follows:

- Reliability growth program's impact on schedule and cost.
- Determination of the reliability growth slot within the framework of the system/product development program.
- The level of the estimated initial reliability without reliability growth testing and its acceptability. Obviously, this information is related to how much of the system/product is new technology.

A manager can use basically two approaches for evaluating the reliability growth process. Both of these approaches are described below [7, 14].

- **Approach I.** This approach monitors various process-related activities for ensuring that such activities are being conducted in a timely manner, and the effort and work quality are according to the program plan. Thus, this approach is activity oriented and is practiced for supplementing the assessments. Sometimes early in the program, it is relied upon totally. The main reason for this is the lack of adequate amount of objective data in the early program stages.
- **Approach II.** This approach uses assessment (i.e., quantitative evaluations of the current reliability status) based on data from the detection of failure sources; thus, it is a results-oriented approach. In order to evaluate reliability growth progress, a comparison between the planned and assessed values is carried out. Subsequently, the appropriate decisions are made.

## 7.7 Reliability growth models

Over the years, many mathematical models concerning reliability growth have been developed [14]. Two such models are presented below.

### 7.7.1 Army material system analysis activity (AMSHA) model

This can be used for tracking reliability growth within test phases [14]. The model allows, for the purpose of reliability growth tracking, the development of rigorous statistical procedures and is based on the following two assumptions:

i. Reliability growth can be modeled as a non-homogeneous Poisson process (NHPP) within the framework of a test phase.
ii. On the basis of failures and test time within the framework of a test phase, the cumulative failure rate is linear on log-log scale.

The following five symbols are associated with this model:

t is the test time from the start of the test phase.
N(t) is the number of system failures by time t.
$\alpha$ and $\theta$ are the parameters.
E[N(t)] is the expected value of N(t).

Thus, the expected value of N(t) is expressed by

$$E\left[N(t)\right] = \theta t^{\alpha} \tag{7.26}$$

By differentiating Equation (7.26) with respect to t, we obtain the following equation for the intensity function:

$$\lambda(t) = \theta\alpha t^{\alpha-1} \tag{7.27}$$

where
$\lambda(t)$ is the intensity function.

The instantaneous MTBF is expressed by

$$n(t) = \frac{1}{\lambda(t)} \tag{7.28}$$

where
n(t) is the instantaneous MTBF.

Additional information on this model is available in Refs. [1, 14].

### 7.7.2 Duane model

This model is basically a graphical approach for performing analysis of reliability growth data and is simple and straight forward to understand. Its two important advantages are as follows [1, 7]:

- Various facts can be clearly depicted by the Duane plot which otherwise could be hidden by statistical analysis. For example, even though the application of a goodness-of-fit test may conclude the rejection of a certain reliability growth model, but it will not clearly provide any possible reasons for the rejection. In contrast, a plot of the same data values might provide some possible reasons for the problem.
- The straight line used by the Duane plot can simply be fitted by eye to the data points.

However, the disadvantages of this model are the reliability parameters that cannot be estimated as well in comparison to a statistical model, and no interval estimates can be calculated.

The cumulative failure rate of this model is expressed by

$$\lambda_c = \frac{F}{T} = \theta T^{-\beta} \tag{7.29}$$

where
$\lambda_c$ is the cumulative failure rate.
T is the total test hours.

F is the number of failures during T.
$\beta$ is a parameter representing the growth rate.
$\theta$ is a parameter determined by circumstances such as design objective, design margin, and product complexity.

In order to estimate the values of the parameters, we take the logarithms of Equation (7.29) to obtain

$$\log \lambda_c = \log \theta - \beta \log T \tag{7.30}$$

It is to be noted that Equation (7.30) is the equation for a straight line. Thus, the plot of the logarithm of the cumulative failure rate, $\lambda_c$, against the logarithm of cumulative operating hours, T, can be used for estimating the values of $\theta$ and $\beta$. The slope of the straight line is equal to $\beta$ and at $T = 1$; $\theta$ is equal to the corresponding cumulative failure rate.

Here, the least-squares method can be employed to have a more accurate straight line fit in estimating the values of $\theta$ and $\beta$.

## 7.8 Problems

1. Discuss the following two classifications of reliability tests:
   - Reliability development and demonstration testing.
   - Qualification and acceptance testing.
2. Assume that 85% reliability of an electronic item is to be demonstrated at 90% confidence level. Calculate the number of electronic items to be placed on test when no failures are allowed.
3. Describe the two methods used to perform an accelerated life test.
4. Describe the reliability growth program.
5. Assume that a sample of 25 identical electronic items was accelerated life tested at 120°C, and their times to failure were exponentially distributed with a mean value of 4000 hours. If the value of the acceleration factor is 15 and the electronic items' normal operating temperature is 30°C, calculate the electronic items, operating at the normal conditions, failure rate, mean time to failure, and reliability for a 3000-hour mission.
6. Describe the reliability growth process evaluation approaches.
7. Describe the following two models:
   - Duane model.
   - AMSAA model.
8. What is the difference between reliability growth modeling and reliability demonstration testing?
9. Assume that 25 identical electronic items were placed on test at time $t = 0$ and none of the failed items were replaced and the test was terminated after 150 hours. Four items failed after 30, 50, 70, and

80 hours of operation. Calculate the electronic items' MTBF and its upper and lower limits with 90% confidence level.

10. Assume that 20 electronic items were put on test at zero time and at the occurrence of the 6th failure, the testing was stopped. The sixth failure occurred at 120 hours, and all the failed items were replaced.

Calculate the MTBF of the electronic items and upper and lower limits MTBF at 80% confidence level.

## *References*


1. MIL-HDBK-781, Reliability Test Methods, Plans and Environments for Engineering Development, Qualification and Production, Department of Defense, Washington, D.C.
2. MIL-STD-781, Reliability Design Qualification and Production Acceptance Test: Exponential Distribution, Department of Defense, Washington, D.C.
3. Mead, P.H., Reliability Growth of Electronic Equipment, Microelectronics and Reliability, Vol. 14, 1975, pp. 439–443.
4. Duane, J.T., Learning Curve Approach to Reliability Monitoring, IEEE Transactions on Aerospace, 1964, pp. 563–566.
5. Dhillon, B.S., Reliability Growth: A Survey, Microelectronics and Reliability, Vol. 20, 1980, pp. 743–751.
6. Van Alven, W.H., Editor, Reliability Engineering, Prentice-Hall, Englewood Cliffs, New Jersey, 1983.
7. Dhillon, B.S., Design Reliability: Fundamentals and Applications, CRC Press, Boca Raton, Florida, 1999.
8. Grant Ireson, W., Coombs, C.F., Moss, R.Y., Handbook of Reliability Engineering and Management, McGraw-Hill, New York, 1996.
9. Nelson, W., Accelerated Testing, John Wiley and Sons, New York, 1980.
10. Bain, L.J., Engelhardt, M., Statistical Analysis of Reliability and Life-Testing Models: Theory, Marcel Dekker, New York, 1991.
11. Meeker, W.Q., Hahn, G.J., How to Plan an Accelerated Life Test: Some Practical Guidelines, American Society for Quality Control (ASQC), Milwaukee, Wisconsin, 1985.
12. Tobias, P.A., Trindade, D., Applied Reliability, Van Nostrand Reinhold, New York, 1986.
13. Elsayed, E.A., Reliability Engineering, Longman, Reading, Massachusetts, 1996.
14. MIL-HDBK-189, Reliability Growth Management, Department of Defense, Washington, D.C.

# chapter eight

# Maintainability management

## 8.1 Introduction

Just like in any other area of engineering, in the practice of maintainability, engineering management plays an important role. The tasks of maintainability management range from simply managing personnel involved with maintainability to effective execution of technical maintainability-related tasks. Thus, maintainability management can be examined from various different perspectives. These perspectives include the place of the maintainability function within the organizational structure, management of maintainability as an engineering discipline, and the role maintainability plays at each phase in the life cycle of system/product under consideration [1, 2].

Some of the early important documents, directly or indirectly, concerning maintainability management are listed as Refs. [3–5]. This chapter presents various important aspects of maintainability management.

## 8.2 Maintainability management functions during the product life cycle

During the product life cycle, as maintainability-related issues arise, various types of maintainability management-related tasks are conducted. An effective maintainability program incorporates a proper dialogue between the user and manufacturer throughout the product life cycle. The product life cycle can be divided into four distinct phases, as shown in Fig. 8.1.

All the phases shown in Fig. 8.1 are described as follows:

- **Concept development phase.** During this phase, the operational needs of the product are translated into a set of operational requirements and high-risk areas highlighted. The primary maintainability management task during this phase is concerned with determining the product effectiveness-related needs, as well as determining, from the product's purpose and intended operation, the required field support policies, and other provisions.

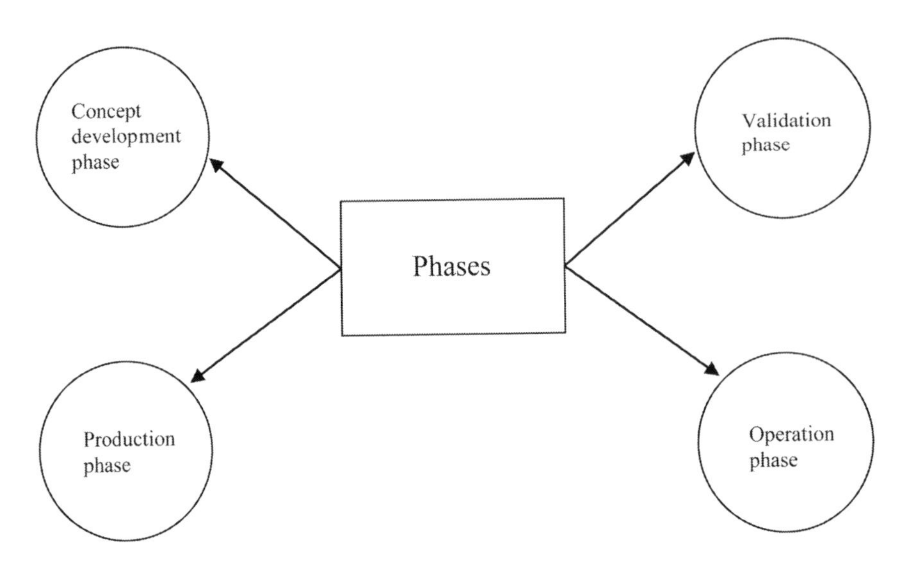


*Figure 8.1* Product life cycle phases.

- **Validation phase.** This is the second phase of product life cycle, and some of the maintainability management tasks associated with phase are as follows:
  - Developing a maintainability program plan that satisfies contractual requirements.
  - Performing maintainability allocations and predictions.
  - Developing a planning document for data collection, analysis, and evaluation.
  - Establishing maintainability incentives and penalties.
  - Participating in design reviews.
  - Developing a plan for maintainability testing and demonstration.
  - Providing appropriate assistance to maintenance engineering in areas such as performing maintenance analysis and developing logistics policies.
  - Establishing maintainability policies and procedures for the validation phase and the subsequent full-scale engineering effort.
  - Coordinating and monitoring maintainability efforts throughout the organization.
- **Production phase.** This is the third phase of the product life cycle, and some of the maintainability management tasks associated with this phase are as follows:
  - Monitoring production-related processes.
  - Evaluating all proposals for changes in regard to their impact on maintainability.

    - Participating in the development of appropriate controls for errors, process variations, and other problems that may affect directly or indirectly maintainability.
    - Ensuring the proper eradication of all shortcomings that may degrade maintainability.
    - Evaluating production test trends from the standpoint of adverse effects on maintainability-related requirements.
- **Operation phase.** This is the fourth and final phase of the product life cycle. Although, no specific maintainability-related tasks are involved with phase, but the phase is probably the most significant because only during this phase, the product's true logistic support and cost-effectiveness are demonstrated. Thus, essential maintainability-associated data can be collected for use in future application.

## *8.3 Maintainability organization functions*

A maintainability organization performs a wide variety of functions. These functions can be grouped under five distinct categories, as shown in Fig. 8.2 [1, 6]. All these categories are described below, separately.

- **Design**
  Maintainability design of a product deals with those features and characteristics of the product that will increase ease of maintenance, make maintenance more cost-effective, and in turn reduce logistic support-related needs. Some of the important design categories functions are as follows:
    - Reviewing product design with respect to maintainability features.
    - Preparing maintainability-related design documents.

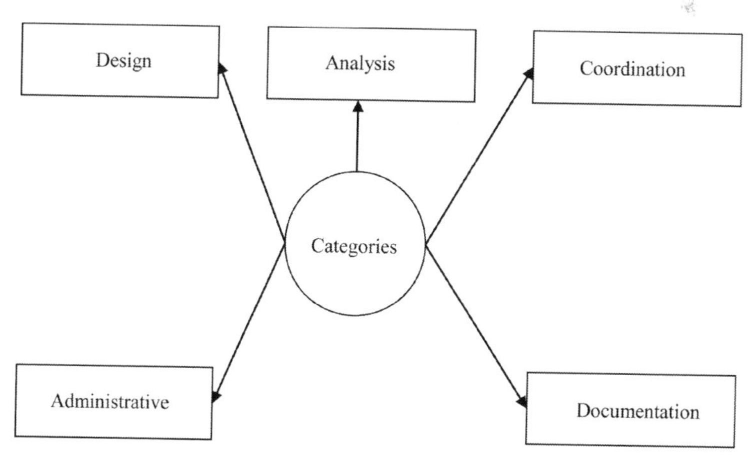


***Figure 8.2*** Categories of maintainability organization functions.

    - Taking part in the development of maintainability design criteria and guidelines.
    - Providing consulting-related services to professionals such as design engineers.
    - Approving design-related drawings from the maintainability standpoint.
- **Analysis**
  A maintainability organization generally spends considerable effort on various analytical-related projects such as maintainability allocation, maintainability prediction, and field data evaluation. Some of the analysis category functions are as follows:
    - Reviewing product specification documents in regard to maintainability-related requirements.
    - Participating in or conducting required maintenance analysis.
    - Performing maintainability allocation and prediction studies.
    - Taking part in product engineering analysis for safeguarding maintainability interests.
    - Conducting analysis of maintainability data obtained from the field and other sources.
    - Preparing maintainability demonstration documents.
- **Coordination**
  Coordination effort is a critical factor in assuring an effective and optimized design, and it accounts for much of the maintainability management effort. Some of the coordination category functions are as follows:
    - Coordinating with bodies such as professional societies, governments, and trade associations on maintainability-related matters.
    - Interfacing with product engineering and other engineering disciplines.
    - Acting as a liaison with subcontractors on maintainability-related matters.
    - Coordinating maintainability training activities for all personnel involved.
- **Documentation**
  The maintainability effort generates and uses a significant amount of information and data. Thus, for the sake of achieving a cost-effective, coherent, and comprehensive design, the efficient and effective handling of this information is very important. Some of the documentation category functions are as follows:
    - Documenting maintainability design review-related results.
    - Documenting the findings of maintainability analysis and tradeoff studies.
    - Establishing and maintaining a library facility that contains all important maintainability documents and information.

    - Establishing and maintaining a maintainability data bank.
    - Developing maintainability-related data and feedback reports.
    - Documenting information related to maintainability management.
    - Preparing and maintaining handbook data and information with respect to maintainability.
- **Administrative**
  The administrative function encompasses those tasks concerned with schedule, cost, and performance, as well as it provides overall direction control to maintainability program management [3]. Some of the administrative category functions are as follows:
    - Organizing the maintainability effort.
    - Preparing schedule and budgets.
    - Developing and issuing policies and procedures for application in maintainability efforts.
    - Acting as a liaison with higher-level management and other concerned bodies.
    - Assigning maintainability-related responsibilities.
    - Providing maintainability training as appropriate.
    - Taking part in program management and design reviews.
    - Monitoring the maintainability organization's output.
    - Preparing a maintainability program plan.

## *8.4 Maintainability program plan*

This is an important document, and it contains maintainability-related information concerning a project under consideration. This document is developed by either the product/system manufacturer or the user, depending on factors such as the philosophy of the involved decision makers and the nature of the project.

Some of the important elements of a maintainability program plan are shown in Fig. 8.3 [7].

All program elements shown in Fig. 8.3 are described below, separately.

- **Objectives:** These are basically the descriptions of the overall requirements for the maintainability program as well as goals of the plan. They also usually include a brief description of the product under consideration.
- **Policies and procedures:** Their main purpose is to assure customers that the group implementing the maintainability program will conduct its assigned task effectively. It is to be noted that under the policies and procedures, the management's overall policy directives for maintainability are also incorporated or referenced. The directives address topics such as maintainability demonstration methods, techniques to be used for maintainability allocation and prediction,

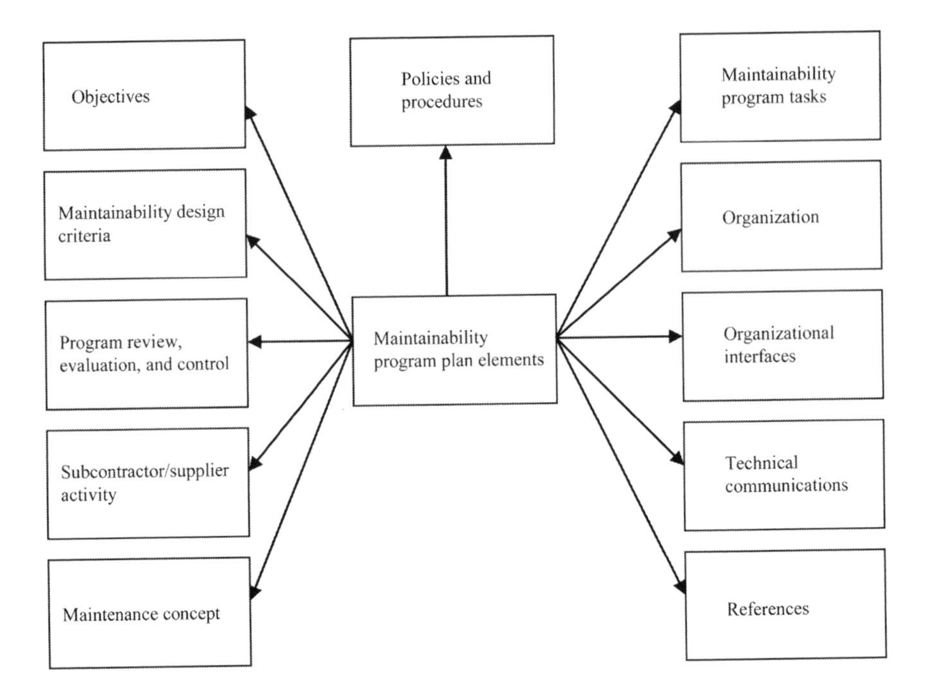


*Figure 8.3* Maintainability program plan elements.

participation in design reviews and evaluation, and data collection and analysis.

- **Maintainability program tasks:** In this section, each program task, task schedule and major milestones, task input requirements, expected task output results, and projected cost are described in detail.
- **Organization:** This section depicts the overall structure of the enterprise as well as provides a detailed organizational breakdown of the maintainability group involved in the project. Furthermore, the section also provides information concerning the background and experience of the maintainability personnel, a work breakdown structure, and a list of personnel assigned to each task.
- **Organizational interfaces:** This section discusses the relationships and lines of communication between the maintainability group and the overall organization. The areas of interface include product engineering design, reliability engineering, testing and evaluation, logistic support, suppliers and customers, and human factors.
- **Maintainability design criteria:** It discusses or references specific maintainability design features applicable to the product under consideration. Furthermore, the description may relate to qualitative and quantitative factors concerning part selection, accessibility, interchangeability, or packaging.

- **Program review, evaluation, and control:** This section describes the methods and techniques to be used for technical design review, program reviews, and feedback control. Furthermore, it describes a risk management plan and discusses the evaluation and incorporation of proposed changes and appropriate corrective measures to be initiated in given situations.
- **Technical communications:** This section briefly describes every deliverable data item and their associated due dates.
- **Subcontractor/supplier activity:** This section discusses the relationships of the organization with suppliers and subcontractors connected to the maintainability program. Furthermore, it outlines the procedures to be used for review and control within those relationships.
- **Maintenance concept:** This section discusses basic maintenance-related requirements of the product under consideration and issues such as qualitative and quantitative objectives for maintenance and maintainability, test and support equipment criteria, organizational responsibilities, spare and repair part factors, and operational and support concepts.
- **References:** This section lists all documents relevant to the maintainability program requirements (e.g., specifications, plans, and applicable standards).

## 8.5 Maintainability design reviews

Design reviews are a critical component of modern design practices, and they are carried out during the product/system design phase. The primary objective of design reviews is to determine the progress of the ongoing design effort and to ensure that correct design principles are being applied. Members of the design review team assess existing and potential problems in various areas concerned with the product under consideration including maintainability.

Nonetheless, design reviews fall into three distinct categories, as shown in Fig. 8.4. These categories are preliminary design review, intermediate design review, and final design review.

The preliminary design review comes prior to formulation of the initial design, and its purpose is the careful examinations of the functions the product must perform and the standards it must meet. The intermediate design review is held prior to developing detailed production drawings, and its purpose is comparing each specification requirement with the proposed design. These requirements may involve items such as maintainability, cost, usage of standard parts, human factors, reliability, safety, performance, and schedule.

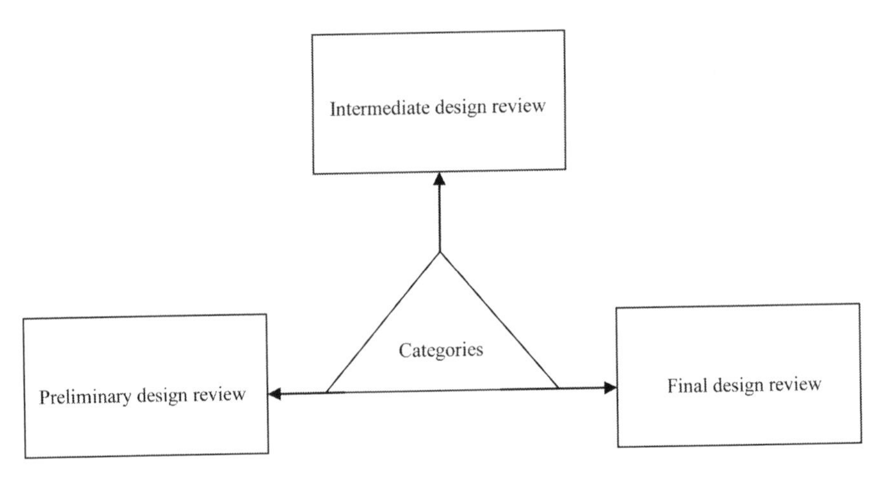


*Figure 8.4* Design review categories.

The final design review generally takes place soon after the completion of production drawings. This review emphasizes factors such as manufacturing methods, analysis results, quality control of incoming parts, design producibility, and value engineering. Additional information on design review categories is available in Refs. [6, 8, 9].

Nonetheless, there are many maintainability-related issues that require careful attention during these design reviews. Some of these issues are as follows [4, 6, 8, 9]:

- Maintainability prediction results.
- Conformance to maintainability design specifications.
- Identified maintainability problem areas and proposed corrective measures.
- Design constraints and specified interfaces.
- Results of maintainability trade-off studies.
- Maintainability demonstration test data.
- Failure mode and effect analysis results.
- Verification of maintainability design test plans.
- Use of built-in monitoring and fault-isolation equipment.
- Maintainability test data obtained from experimental models and breadboards.
- Use of on-line repair with redundancy.
- Use of automatic test equipment.
- Maintainability assessment using test data.
- Use of unit-replacement approach.
- Physical configuration and layout drawings and schematic designs.
- Assessments of maintenance and supportability.

- Corrective measures taken and proposed.
- Selection of parts and materials.
- Proposed changes to the maintenance concept.

## 8.6 Maintainability-associated personnel

Beside maintainability engineers, there are many other engineering professionals who directly or indirectly contribute to system/product maintainability. These professionals include reliability engineers, safety engineers, human factors engineers, and quality control engineers. This section discusses the functions of a maintainability engineer, reliability engineer, safety engineer, human factors engineer, and quality control engineer, respectively [6].

### 8.6.1 Maintainability engineer

This individual plays a very important role in system/product design and has often served before as a design engineer. Some of the tasks performed by a maintainability engineer are maintainability allocation, performing analysis of maintainability feedback data, maintainability prediction, participating in the development of maintainability design criteria, developing maintainability demonstration documents, preparing maintainability design documents, and reviewing product design with respect to maintainability features.

### 8.6.2 Reliability engineer

As is the case with maintainability engineer, this person has also often served as a design engineer. Reliability engineer assists management personnel in defining, evaluating, and containing risks. The responsibilities of a reliability engineer include participating in design reviews, reliability allocation, monitoring subcontractor's reliability programs, analyzing customer requirements, reliability prediction, developing reliability test and demonstration procedures, assessing the effect of environment on product/system reliability, and developing reliability growth monitoring procedures and reliability evaluation models.

### 8.6.3 Safety engineer

During the design phase, the role played by this professional is very important as the newly designed system/product must be safe to operate and maintain. Some of the tasks of a safety engineer are analyzing new designs from the safety standpoint; analyzing historical data on product/system hazards, failures, and accidents; establishing criteria for analyzing any accident associated with a product/system manufactured by the company; monitoring subcontractors' safety efforts; developing safety warning devices; and keeping management continually informed about safety program performance.

### 8.6.4 Human factors engineer

The maintainability of a system/product will depend on a large degree of decisions made by the human factors engineer. Therefore, this person should be trained in fields such as physiology, anthropometry, and/or psychology [6, 10]. He/she determines the tasks required for making full use of the product; how to label control devices most clearly; the most appropriate way of grouping the various tasks required into individual jobs; the arrangement of control devices that generates optimum use; the environmental conditions that effect individual performance and physical well-being; the best way of displaying specified information, and of arranging visual displays for ensuring optimum use, especially when the user must split his/her attention between two or more displays; how human decision-making and adaptive abilities can be put to the best use; what level of information flow or required decision making overburdens operators; and other similar matters [6, 11].

### 8.6.5 Quality control engineer

Although, this person does not necessarily belong to one of the engineering disciplines, but he/she plays an important role during the product life cycle, particularly during the design and manufacture phases. The quality control engineer develops and applies appropriate inspection-related plans, uses statistical quality control methods, evaluates the quality of procured parts, participates in quality-associated meetings, analyzes quality-related defects, audits the quality control system at appropriate times, and develops quality related standards [6, 12].

## 8.7 Problems

1. Discuss maintainability management-related tasks during the following two phases of the product life cycle:
    - Validation phase.
    - Production phase.
2. Discuss the following two categories of maintainability organization functions:
    - Design.
    - Analysis.

3. What is a maintainability program plan?
4. List at least ten important elements of a maintainability program plan.
5. Describe the following three elements of a maintainability program plan:
   - Organizational interfaces.
   - Policies and procedures.
   - Maintainability design criteria.
6. Describe maintainability design reviews.
7. Discuss the functions of the following two professionals:
   - Maintainability engineer.
   - Safety engineer.
8. List at least ten maintainability-related issues that require careful attention during design reviews.
9. Discuss the functions of the following two professionals:
   - Human factors engineer.
   - Reliability engineer.
10. Define the term maintainability management.

## References

1. AMCP-706-133, Engineering Design Handbook: Maintainability Engineering Theory and Practice, Department of Defense, Washington, D.C., 1976.
2. Dhillon, B.S., Reiche, H., Reliability and Maintainability Management, Van Nostrand Reinhold, New York, 1985.
3. MIL-STD-470, Maintainability Program Requirements for Systems and Equipment, Department of Defense, Washington, D.C., 1966.
4. Patton, J.D., Maintainability and Maintenance Management, Instrument Society of America, Research Triangle Park, North Carolina, 1980.
5. Blanchard, B.S., Lowery, E.E., Maintainability, McGraw-Hill, New York, 1969.
6. Dhillon, B.S., Engineering Maintainability, Gulf Publishing, Houston, Texas, 1999.
7. Blanchard, B.S., Verma, D., Peterson, E.L., Maintainability: A Key to Effective Serviceability and Maintenance Management, John Wiley and Sons, New York, 1995.
8. AMCP-706-134, Engineering Design Handbook: Maintainability Guide for Design, Department of Defense, Washington, D.C., 1972.
9. Pecht, M., Editor, Product Reliability, Maintainability, and Supportability Handbook, CRC Press, Boca Raton, Florida, 1995.
10. Woodson, W.E., Human Factors Design Handbook, McGraw-Hill, New York, 1981.
11. McCormick, E.J., Human Factors Engineering, McGraw-Hill, New York, 1970.
12. Hayes, G.E., Romig, H.G., Modern Quality Control, Collier MacMillan Publishers, London, 1977.

# chapter nine

# Human factors in maintainability

## 9.1 Introduction

Human factors is a quite important field of engineering, and it exists because humans make errors in using and maintaining engineering systems/equipment; otherwise, it would be rather difficult to justify the field's existence. Over the years, the terms human factors, ergonomics, human factors engineering, and human engineering in the published literature have appeared interchangeably. Nonetheless, human factors are a body of scientific facts concerning human characteristics (the term includes all biomedical and psychosocial considerations).

Although the modern history of human factors may be traced back to 1898, when Frederick W. Taylor performed various studies to determine the most suitable design of shovels; human factors have only been an important component of maintainability work since World War II [1–3]. During this war, the military equipment's performance clearly proved that equipment is only as good as the persons operating and maintaining it. This simply means that humans play a very important role in the overall success of a system/equipment. Systems may malfunction for various reasons including poor attention given to human factors in regard to maintainability during the design process [2].

This chapter presents various important aspects of human factors in maintainability.

## 9.2 General human behaviors

Over the years, researchers around the world have studied human behaviors and made conclusions about many general, expected, and typical behaviors. In maintainability work, good knowledge of such behaviors can be very useful directly or indirectly. Some general human behaviors are presented below [3–5].

- Humans are reluctant to admit mistakes or errors.
- Humans get easily confused with unfamiliar things.
- Humans have tendency to use their hands for testing or examining.
- Humans regard manufactured products/items as being safe.

- Humans have become accustomed to certain color meanings.
- Humans often estimate speed or clearance poorly.
- Humans, in general, have a very little idea about their physical limitations.
- Humans conduct their tasks while thinking about other things.
- In emergencies, humans generally respond irrationally.
- Humans can get easily distracted by certain aspects of a product's features.
- Humans read instructions and labels incorrectly or overlook them altogether.
- Humans fail to recheck their work for errors after performing a procedure.
- Humans are too impatient to take the appropriate amount of time for observing precautions.
- Humans become complacent and less careful after successfully handling hazardous items over a lengthy time period.
- Humans assume that an object is small enough to get hold of and is light enough to pick up.
- Humans' attention is drawn to items such as loud noises, flashing lights, bright and vivid colors, and bright lights.
- Humans generally overestimate short distances and underestimate large and horizontal distances.
- Humans are rather reluctant to admit that they do not see objects clearly, whether because of poor eyesight or inadequate illumination.
- Humans expect electrical switches to move upward or to the right for turning power on.
- Humans generally expect the valve handles and faucets rotate counter clockwise for increasing the flow of liquid, stream, or gas.

## 9.3 Human body measurements

This information is very important in designing for maintainability since humans generally operate and maintain engineering systems/products. It is very helpful to designers for ensuring that equipment and products under consideration will properly accommodate operating and maintenance personnel of varying sizes, shapes, and weights. In turn, these individuals will conduct their tasks effectively. Generally, human body-associated requirements are stated in the system/product design specification, particularly, when the equipment is being developed for use by the military personnel. For example, MIL-STD-1472 [6] states 'Design shall insure operability and maintainability by at least 90 percent of the user

*Table 9.1* Some body-associated dimensions of the U.S. adult population ranging from 18 to 79 years of age

| | | 5th percentile (in inches) | | 95th percentile (in inches) | |
|---|---|---|---|---|---|
| No. | Description | Male | Female | Male | Female |
| 1 | Standing height | 63.6 | 59 | 72.8 | 67.1 |
| 2 | Seated width | 12.2 | 12.3 | 15.9 | 17.1 |
| 3 | Seated eye height | 28.4 | 27.4 | 33.5 | 31.0 |
| 4 | Sitting height | 33.2 | 30.9 | 38.0 | 35.7 |
| 5 | Weight | 126 (lb) | 104 (lb) | 217 (lb) | 199 (lb) |

population' and 'The design range shall include at least the 5th and 95th percentiles for design-critical body dimensions'.

Furthermore, the standard also states that the anthropometric data's use should take into consideration factors such as follows:

- The position of the body during task performance.
- The increments in the design-critical dimensions imposed by protective garments.
- The nature and frequency of tasks to be conducted.
- The mobility and flexibility requirements of the task.
- The need to compensate for obstacles.
- The difficulties associated with intended tasks.

Some body-associated dimensions of the U.S. adult population ranging from 18 to 79 years of age are presented in Table 9.1 [5–7].

Some of the useful pointers considered quite useful for engineering designers, concerning the application of body strength and force, are as follows [3, 8]:

- The maximum push force for the side-to-side motion is around 90 pounds.
- An individual's arm strength reaches its peak about age 25.
- The degree of force that can be exerted is determined by various factors including body position, the object involved, direction of force applied, and body parts involved.
- With the use of the whole arm and shoulder, the maximum exertable force is increased.
- Pull force is higher from a sitting than from a standing position.
- The maximum handgrip strength of a 25-year-old male is around 125 pounds.

## 9.4 Human sensory capabilities

In maintainability-related work, there is a definite need for a good understanding of human sensory capacities as they apply to areas such as noise, color coding, parts identification, and shape coding. The five major senses possessed by humans are sight, hearing, touch, smell, and taste. In addition, human can sense items such as temperature, pressure, vibration, acceleration (shock), and linear motion.

The first three of the five major senses are described below, and additional information on other senses is available in Refs. [2, 3, 9].

### 9.4.1 Sight

This sensor plays an important role in maintainability work and is stimulated by electromagnetic radiation of certain wavelengths, often referred to as the visible segment of the electromagnetic spectrum. In daylight, the eyes of humans are very sensitive to greenish-yellow light, and they see differently from different angles.

Some of the important factors concerning color in regard to the human eye are as follows:

- The color reversal phenomenon may take place when one is staring, for example, at a green or red light and then glances away. In such circumstances, the signal to the brain may reverse the color.
- Generally, the eye can perceive all colors when looking straight ahead. However, with an increase in viewing angle, color perception decreases quite significantly.
- In poorly lit areas or at night, it may be impossible to determine correctly the color of a small point source of light (e.g., a small warning light) at a distance. In fact in such situations, the light colors will appear to be white.

Some guidelines considered quite useful for designers and others are to select colors in such a way that color-weak individuals do not get confused, make use of red filters with a wavelength greater than 6500 $\overset{0}{A}$, and avoid placing too much reliance on color when critical tasks are to be conducted by fatigued people [10].

### 9.4.2 Hearing

This sensor can also be an important factor in maintainability work, as excessive noise may lead to problems such as reduction in the workers' efficiency, loss in hearing if exposed for long periods, adverse effects on tasks, and need for intense concentration or a high degree of muscular coordination. Thus, for reducing the effects of noise, some useful guidelines directly or indirectly related to maintainability are as follows [2]:

- Properly incorporate into the equipment appropriate acoustical design and mufflers and other sound-proofing devices in areas where maintenance-related tasks must be conducted in the presence of extreme noise.
- Prevent all unprotected maintenance/repair personnel from entering areas with sound levels greater than 150 dB.
- Protect all maintenance/repair personnel by issuing protective devices where noise reduction is not possible.
- Keep noise levels below 85 dB in areas where the presence of maintenance personnel is absolutely necessary.

### 9.4.3 Touch

This sensor compliments human ability to interpret auditory and visual stimuli. In maintenance-related activities, the touch sensor may be used for relieving eyes and ears of part of the load. For example, its usage could be the recognition of control knob shapes with or without using other sensors.

It is to be noted that the use of the touch sensor in technical-related work is not new; it has been used for many centuries by craft workers to detect surface roughness and irregularities. Furthermore, as per Ref. [11], the surface irregularities' detection accuracy dramatically improves when the involved worker moves an intermediate piece of paper or thin cloth over the surface of object rather than simply using his/her bare fingers.

## 9.5 Visual and auditory warning devices in maintenance activities

For the safety of maintenance personnel, in maintenance work, various visual and auditory warning devices are used. A clear understanding of such devices is absolutely essential. Some examples of warning devices used in maintenance work are bells, buzzers, and sirens. Thus, in maintainability design, in regard to the use of auditory warning devices attention should be given to factors such as follows [3]:

- Easy detectability.
- Suitability to get the attention of repair personnel.
- Noncontinuous and high-pitched tones above 2000 Hz.
- Use of warbling or undulating tones and sound at least 20 dB above threshold level.

- Distinctiveness.
- No requirement for interpretation when maintenance personnel are performing repetitive tasks.

Additional design-related recommendations for auditory alarm and warning devices for addressing corresponding conditions (in parentheses) modulate the signal to generate intermittent beeps (signal must command maintenance person's attention), select a frequency that makes the signal audible through other noise (presence of background noise), use manual shut-off mechanism (warning signal must be acknowledged), use high intensities and avoid high frequencies (repair personnel are performing their tasks far from the signal source), and use low frequencies (sound is expected to pass through partitions and bend around obstacles).

Some of the conditions for using visual presentation are that the maintenance person is overburdened with auditory stimuli, the message does not require immediate action, the message is quite complex, the maintenance person's job allows him/her to remain in one place, the message receiving location is too noisy, and the message is long. Similarly, some of the conditions for using auditory presentation are that the maintenance person is moving around continuously, the message requires immediate action, the maintenance person is overburdened with stimuli, the message receiving location is too brightly lit, the message is short, and the message is simple.

Additional information on conditions for using visual and auditory presentations is available in Ref. [3].

Finally, it is added that there are also situations that require the simultaneous use of both visual and auditory signals. These situations are shown in Fig. 9.1 [2, 3]. An example of a situation: environmental condition is high noise level/poor illumination that prevents data presentation through either visual or auditory means alone.

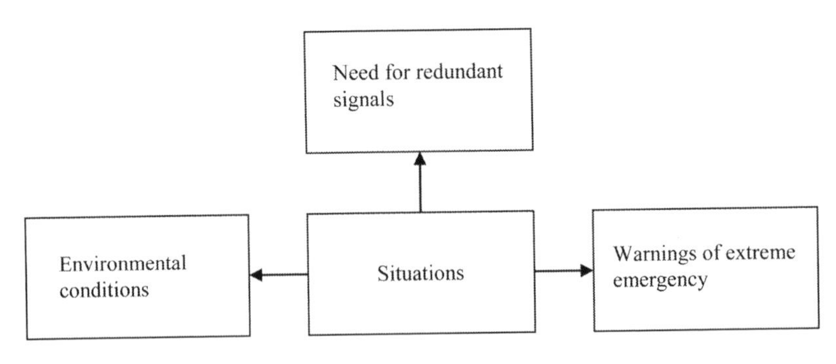


*Figure 9.1* Situations that require the simultaneous use of both auditory and visual signals.

## 9.6 Human factors formulas

Over the years, researchers in the area of human factors have developed many mathematical formulas to estimate human factors-related information. This section presents some of the formulas considered useful for application in the area of maintainability engineering.

### 9.6.1 Character height estimation formulas

#### 9.6.1.1 Formula I

This formula was developed in 1959 by Peters and Adams [12] for determining character height by taking into consideration factors such as viewing distance, illumination, viewing conditions, and importance of reading accuracy. Thus, the character height is defined by

$$H_c = \beta V_d + \theta_i + \theta_v \tag{9.1}$$

where

$H_c$ is the character height in inches.
$V_d$ is the viewing distance expressed in inches.
$\beta$ is a constant whose specified value is 0.0022.
$\theta_i$ is the correction factor associated with importance, and its stated value for items such as emergency labels is 0.075 and for other items is $\theta_i = 0$.
$\theta_v$ is the correction factor for viewing conditions and illumination. Its specified values for various corresponding viewing conditions and illuminations (in parentheses) are 0.26 (unfavourable reading conditions, below 1-foot candle), 0.16 (favourable reading conditions, below 1-foot candle), 0.16 (unfavourable reading conditions, above 1-foot candle), and 0.06 (favourable reading conditions, above 1-foot candle).

**Example 9.1**

Assume that the viewing distance of an instrument is estimated to be 50 inches. Calculate the height of the characters that should be used on the panel for $\theta_i = 0.075$ and $\theta_v = 0.26$.

By using the specified data values in Equation (9.1), we obtain

$$H_c = (0.0022)(50) + 0.075 + 0.26 = 0.445$$

Thus, the height of the characters should be 0.445 inches.

#### 9.6.1.2 Formula II

Generally, for a comfortable arm reach to conduct control and adjustment-oriented tasks, the instrument panels are installed at a viewing distance of 28 inches. Thus, number, marking, and letter sizes are based on this viewing distance. However, sometimes the need may arise to change this distance; under such circumstances, the following equation can be utilized for estimating the required character height [13, 14]:

$$H_c = H_{st} D_v / 28 \tag{9.2}$$

where

$H_c$ is the character height estimate at the stated viewing distance, $D_v$, expressed in inches.

$H_{st}$ is the standard/recommended character height at a viewing distance of 28 inches.

**Example 9.2**

Assume that a meter has to be read at a distance of 40 inches. The standard numeral height at a viewing distance of 28 inches at low luminance is 0.31 inches. Calculate the numeral height for the viewing distance of 40 inches.

By inserting the stated data values into Equation (9.2), we get

$$H_c = (0.31)(40)/28 = 0.44 \text{ inches}$$

Thus, the estimate for the numeral height for the stated viewing distance is 0.44 inches.

### 9.6.2 Glare constant estimation

For various maintainability-associated tasks, glare can be a serious problem. The value of the glare constant can be estimated by using the following equation [15]:

$$\theta_g = \frac{(SL)^{1.6}(SA)^{0.8}}{(ABG)^2(GBL)} \tag{9.3}$$

where

$\theta_g$ is the glare constant value.

SL is the source luminance.

SA is the solid angle subtended at the eye by the source.

ABG is the angle between the viewing direction and the glare source direction.

GBL is the general background luminance.

### 9.6.3 The decibel

This basic unit of sound/noise intensity is named after Alexander Graham Bell (1847–1922), the inventor of the telephone. The sound-pressure level is expressed by [14, 16]

$$SPL = 10\log_{10}\left[\frac{SP^2}{P_f^2}\right] \tag{9.4}$$

where

SPL is the sound-pressure level expressed in decibels.
$SP^2$ is the sound pressure, squared, of the sound to be measured.
$P_f^2$ is the standard reference sound pressure squared, representing zero decibels. Under normal circumstances, $P_f$ is the faintest 1000 Hz tone that an average young person can hear.

### 9.6.4 Lifting load estimation

This formula is concerned with calculating the maximum lifting load for an individual. This information could very useful in regard to structuring various maintenance-related tasks. The maximum lifting load is defined by

$$MLL_p = (IBMS_p)c \tag{9.5}$$

where

$MLL_p$ is the maximum lifting load for a person.
$IBMS_p$ is the isometric back muscle strength of the person.
c is a constant whose values are 0.95 and 1.1 for females and males, respectively.

## 9.7 Problems

1. List at least 20 general human behaviors.
2. Write an essay on human factors in maintainability.
3. List at least six useful pointers, considered quite useful, for engineering designers concerning the application of body strength and force.
4. What are the major senses possessed by humans?
5. Discuss the following two human sensory capabilities:
   - Touch.
   - Sight.
6. List at least six factors to which attention should be given in maintainability design with respect to the use of auditory warning devices.
7. What are the situations that require the simultaneous use of both visual and auditory signals?

8. Assume that the viewing distance of an instrument panel is estimated to be 40 inches. Calculate the height of the characters that should be used on the panel if the values of the importance correction factor and the viewing conditions and illumination correction factor are 0.075 and 0.26, respectively.
9. Write down the formula to estimate the maximum lifting load for a person.
10. Write down the formula for estimating the glare constant.

## References

1. Chapanis, A., Man-Machine Engineering, Wadsworth Publishing, Belmont, California, 1965.
2. AMCP-706-134, Engineering Design Handbook: Maintainability Guide for Design, Department of Defense, Washington, D.C., 1972.
3. Dhillon, B.S., Engineering Maintainability, Gulf Publishing, Houston, Texas, 1999.
4. Nertney, R.J., Bullock, M.G., Human Factors in Design, Report No. ERDA-76-45-2, The Energy Research and Development Administration, U.S. Department of Energy, Washington, D.C., 1976.
5. Woodson, W.E., Human Factors Design Handbook, McGraw-Hill, New York, 1981.
6. MIL-STD-1472, Human Engineering Design for Military Systems, Equipment, and Facilities, Department of Defense, Washington, D.C., 1972.
7. Dhillon, B.S., Advanced Design Concepts for Engineers, Technomic Publishing, Lancaster, PA, 1998.
8. Henney, K., Editor, Reliability Factors for Ground Electronic Equipment, The Rome Air Development Center, Griffiss Air Force Base, Rome, New York, 1955.
9. AMCP 706-133, Engineering Design Handbook: Maintainability Engineering Theory and Practice, Department of Defense, Washington, D.C., 1976.
10. Woodson, W., Human Engineering Suggestions for Designers of Electronic Equipment, in NEL Reliability Design Handbook, U.S. Naval Electronics Laboratory, San Diego, California, 1955, pp. 12.1–12.5.
11. Lederman, S., Heightening Tactile Impressions of Surface Texture, in Active Touch, edited by G. Gordon, Pergamon Press, Elmsford, New York, 1978, pp. 20–32.
12. Peters, G.A., Adams, B.B., Three Criteria for Readable Panel Markings, Product Engineering, Vol. 30, No. 2, 1959, pp. 55–57.
13. Dale Huchingson, R., New Horizons for Human Factors in Design, McGraw-Hill, New York, 1981.
14. McCormick, E.J., Sanders, M.S., Human Factors in Engineering and Design, McGraw-Hill, New York, 1982.
15. Oborne, D.J., Ergonomics at Work, John Wiley and Sons, New York, 1982.
16. Adams, J.A., Human Factors Engineering, MacMillan, New York, 1989.
17. Poulsen, E., Jorgensen, C., Back Muscle Strength, Lifting and Stooped Working Postures, Applied Ergonomics, Vol. 2, 1971, pp. 133–137.

# chapter ten

# Maintainability testing and demonstration

## 10.1 Introduction

The primary function of maintainability testing and demonstration is verifying the maintainability–related features that have designed and built into a system/product [1, 2]. Maintainability testing and demonstration also provide the customers with confidence, prior to making any production commitments, that the system/equipment design under consideration meets the maintainability-related requirements. Prior to the testing and demonstration phase, the maintainability program tasks have been basically analytical. The main drawback of the analytical evaluations is that they do not reflect practical experience with the actual hardware.

Thus, it is quite essential to add realistic evaluations to analytical evaluations by performing real maintainability tests and demonstrations with the equipment/system in its operational environment.

This chapter presents various important aspects of maintainability testing and demonstration.

## 10.2 Maintainability testing and demonstration planning and control requirements

In order to gain maximum benefits from maintainability, tests and demonstrations require good planning and control. Thus, maintainability testing and demonstration planning and control requirements may be categorized under six classifications, as shown in Fig. 10.1 [2, 3].

All the six classifications shown in Fig. 10.1 are described as follows.

### 10.2.1 Creating a demonstration model

This is basically concerned with developing/creating a mock-up model representing the final product to demonstrate maintainability features. A mock-up model serves the following two basic functions [2]:

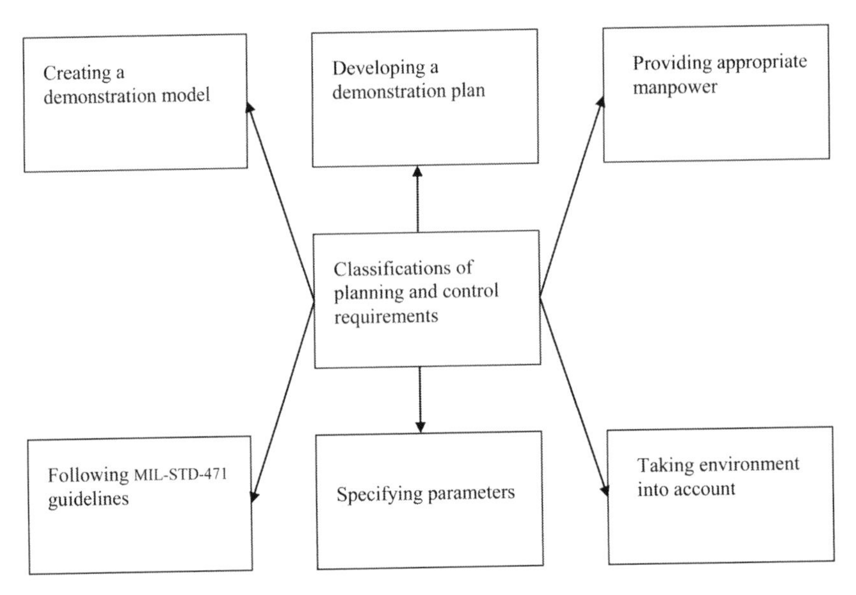


***Figure 10.1*** Classifications of maintainability testing and demonstration planning and control requirements.

- Providing a basic mechanism to demonstrate the product's/system's proposed quantitative parameters and qualitative design features for maintainability.
- Providing a designer's tool for visibility, packaging limitation, experimentation, and planning, prior to release of the final design/drawing.

### *10.2.2 Developing a demonstration plan*

A good demonstration plan should cover areas such as test planning, administration, and control; test conditions; and test documentation, analysis, and report. Also, it should conform to the specifics described in Ref. [4].

During the early manufacturer/contractor participation in a program such as the validation phase, the first step is conceiving, proposing, and negotiating the demonstration test planning subject. As the program progresses, the mutually agreed-upon test plans are updated in regard to items such as schedules, demonstration model designation, personnel selection, and identification of logistic support resource-related requirements. The key factor in accomplishing the demonstration on schedule and within budget is the demonstration's administration and control. Some elements of administration and control are as follows:

- Method of organization.
- Test monitoring.
- Test event scheduling.
- A team approach, if desired.
- Assignment of responsibilities.
- Cost control.
- Organizational interfaces.
- Test data collection, reporting, and analysis.

It is to be noted that the complexity and type of the equipment/system under consideration plays a key role in shaping the requirements for test documentation. Documentation requirements generally include items such as failure reports, demonstration task data sheets, demonstration work sheets, task selection work sheets, demonstration analysis work sheets, frequency and distribution work sheets, interim demonstration reports, and final reports.

### 10.2.3 *Following MIL-STD-471 guidelines*

This document developed by the U.S. Department of Defense lays out guidelines that equipment/system manufacturers should consider in the planning and control of maintainability demonstrations [4]. The document covers the following topics [4]:

- Administration, control, reporting, evaluation, and analysis procedures for the demonstration.
- The pre-demonstration, formal demonstration, and phases.
- Test conditions, selecting a test method, establishing test teams, and suggested test support materials.
- Data collection.
- Selection, performance, and sampling of corrective and preventive maintenance tasks.

Finally, it is added that this document (i.e., MIL-STD-471) guidelines are considered essential for an effective maintainability demonstration.

### 10.2.4 *Providing appropriate manpower*

The personnel who conduct maintainability demonstrations will be essential to their success. Thus, it is very important that they possess backgrounds and skill levels similar to those of the system's/product's final user, operating, and maintenance personnel. One way to do this is to have such individuals from the client organization to conduct the test.

Useful guidelines for selecting demonstrators are available in Ref. [3].

### 10.2.5 Specifying parameters

The basic purpose of a formal maintainability demonstration process is to verify compliance with stated parameters. Thus, the specifications for demonstration parameters should be stated in quantitative terms. Some examples of measurable time parameters are mean preventive maintenance time (MPMT), mean corrective maintenance time (MCMT), and mean time to repair (MTTR).

### 10.2.6 Taking environment into account

Past experiences over the years have shown clearly that system/equipment downtime may vary quite significantly between laboratory-controlled conditions and actual operational conditions. Thus, environment is an important factor in testing, and those responsible must carefully consider factors such as support resource needs, test facilities, and limitations simulations.

## 10.3 Useful checklists for maintainability demonstration plans, procedures, and reports

In maintainability demonstration, checklists play a very important role. The checklist for maintainability demonstration plans and procedures should cover items such as follows [2, 5]:

- **Purpose and scope:** This is a statement of general test objectives and a general description of the test to be conducted.
- **Test facilities:** This is information such as a description of the test item's configuration, a general description of test facility, identification of the test location, test area security measures, and test safety features.
- **Test requirements:** These requirements include items such as the method of generating a candidate fault list, the method of selecting and applying faults from the candidate list, the levels of maintenance to be demonstrated, a list and schedule of test reports to be issued, and support material requirements.
- **Test participation:** Decisions to be made regarding test participation are the test team members, their assignments, and test decision-making authority.
- **Test monitoring:** This is the method of monitoring and recording test results.
- **Test schedule:** This should include these items: the starting date, the finish date, and the test program review schedule.

- **Test conditions:** Two components of these conditions are the modes of equipment/system operation during testing and a description of the environmental conditions under which the test to be conducted.
- **Test ground rules:** Under this should be a list of items to which the rules apply. These items include maintenance time limits, maintenance inspection, maintenance due to secondary failures, instrumentation failures, and technical manual usage and adequacy.
- **Testability demonstration considerations:** The components of these considerations include the method of selecting and simulating candidate faults, the built-in test requirements to be demonstrated, the repair levels for which requirements will be demonstrated, and acceptable levels of ambiguity at each repair level.
- **Reference documents:** The checklist should also detail all applicable reference documents.

The maintainability demonstration reports checklist include items related to test results, such as maintenance tasks planned, maintenance tasks selected, measured repair times, data analysis calculation, the documentation used during maintenance, the selection method, application of the accept/reject criteria, a discussion of deficiencies highlighted during testing, and qualifications of the personnel conducting tasks.

## 10.4 Maintainability test approaches

All maintainability tests are not formal accept/reject demonstration tests. There are many points in the product/system life cycle and in related maintainability program tasks that need test data, both prior to and after the formal decision to accept or reject. Test data may be necessary for administrative and logistic control for updating corrective actions or modifications, to make decisions regarding maintainability design needs, or for evaluating life cycle maintenance support.

The maintainability test approaches that can provide this type of data fall into six categories, as shown in Fig. 10.2 [2].

All the six test categories of maintainability test approaches shown in Fig. 10.2 are described below.

- **Dynamic tests:** These tests simulate typical operation or uses of system/equipment so that each and every item involved can be checked. The tests involve a continuous input signal and analysis of the corresponding output signals for determining whether system/equipment needs are fully satisfied. Furthermore, dynamic tests also provide additional information on matters such as integration rates, phase characteristics, and frequency responses.

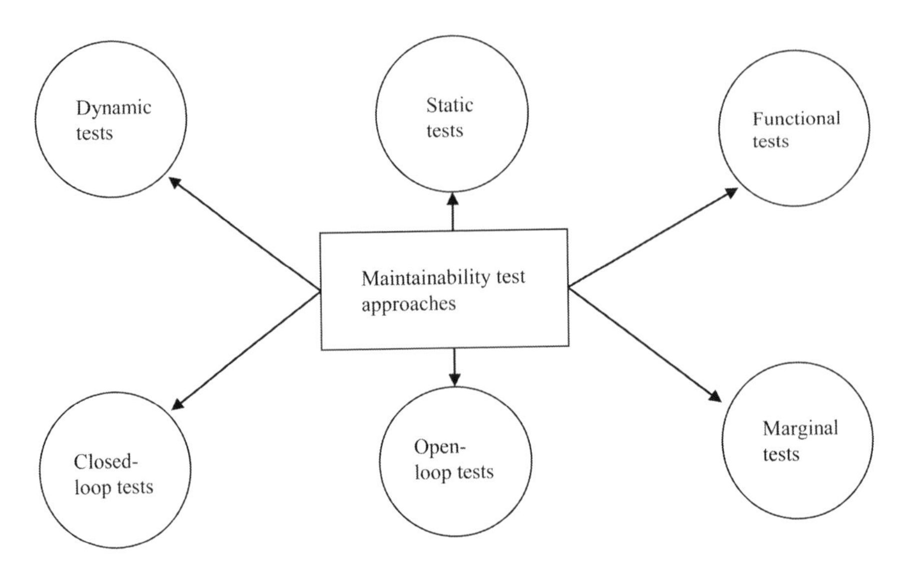


*Figure 10.2* Categories of maintainability test approaches.

- **Static tests:** These tests are simple and straightforward to conduct and provide quite useful information on the transient behavior of the item being tested. A series of intermittent, sequenced input signals feed into the item, and, for measuring its operation, the test monitors all output response signals. Furthermore, it is added that static tests generally establish a confidence factor but their application does not go beyond that.
- **Functional tests:** These tests closely simulate normal operating conditions for establishing the product's state of readiness for conducting its proposed mission effectively. The tests can proceed on a system-wide level or focus on items such as replaceable subassemblies. Finally, it is to be noted that functional tests are conducted and required at each point of the evaluation for the product.
- **Closed-loop tests:** These tests generate information for application in evaluating design effectiveness, tolerance adequacy, performance, and other key issues. In these tests, the stimulus is adjusted continuously according to the performance of the system/equipment under the test conditions. Closed-loop tests are very useful in situations when a high degree of accuracy is required as well as when test points radiate performance-degrading noise levels. However, it is to be noted that closed-loop circuits or paths in system or equipment design are quite difficult to maintain, and failures in the loop are quite difficult to diagnose.

- **Open-loop tests:** These tests represent a refinement of dynamic and static tests, and they do not provide intelligence feedback to the item being tested, that is, the stimulus is not adjusted. Open-loop tests generally provide better maintenance-related information than closed-loop tests because they make a direct observation of the system transfer function without the modifying feedback's influence. This approach is also cheaper and simpler than closed-loop testing and is probably the most appropriate type of testing for maintenance purposes because it eliminates the possibility of test instability.
- **Marginal tests:** The objective of applying these tests is to isolate potential problems through the abnormal operating conditions' simulation. These tests supply unrelated stimulus to the system/equipment under conditions such as extreme heat, vibrations, and lowered power supply voltages. Furthermore, these tests provide greatest value as part of fault prediction, where they highlight various incipient failures resulting from abnormal operating conditions and environments.

## 10.5 Maintainability testing methods

As a single maintainability parameter can seldom address all desirable maintainability characteristics, there are many different test methods to address many diverse maintainability parameters. Eleven of these methods are presented below [2, 4, 6]. It is to be noted that Ref. [4] provides the statistical aspect of the various methods presented below.

### 10.5.1 Method I: The mean method

This method is considered quite useful when the need/requirement is stated in terms of a mean value as well as there is a corresponding design goal value. The test plan is composed of Test Plan A and Test Plan B. Plan A assumes lognormal distribution to determine the sample size. It also assumes that a lognormal distribution can satisfactorily represent the maintenance times and that the variances of the logarithms of the maintenance times are already stated. Test plan B makes no such assumptions. However, on the other hand, both are fixed sample tests, and a minimum sample size of 20, that make use of the Central Limit Theorem and the asymptotic normality of the sample mean in their development.

### 10.5.2 Method II: The critical maintenance time or man-hours method

This method is applicable when the need is stated in terms of the following two factors [2, 6]:

- **Factor I:** Required critical maintenance time or critical man-hours.
- **Factor II:** A corresponding design goal value.

This test is distribution-free and can be used for establishing a critical upper limit on the time or man-hours needed to conduct certain maintenance tasks. The following factors are associated with this test method:

- There is no need for assuming the distribution of maintenance time or man-hours.
- Both the alternate and null hypotheses refer to a fixed time and the percentile varies.

### 10.5.3 Method III: The mean maintenance time and maximum maintenance time method

This method demonstrates maintainability indices such as follows:

- Mean maintenance time that includes both corrective and preventive actions.
- MCMT.
- MPMT.

For demonstrating MCMT, the procedures of this method are based on the Central Limit Theorem. As the information on the variance of maintenance times is not required, this allows the method to be used with any underlying distribution, provided the sample size is at least 30. Finally, it is added that the maximum maintenance time demonstration/procedure is valid for cases with lognormal underlying distribution of corrective maintenance task times.

### 10.5.4 Method IV: The critical percentile method

Maintainability demonstrations employ this method when the requirement is expressed in terms of:

- A required critical percentile
- A corresponding design goal value

It is to be noted that if the critical percentile is fixed at 50%, then this method is referred to as the test of a median. The basis for the decision criteria is the asymptotic normality of the maximum likelihood estimate of the percentile value. The method is based on the following two assumptions:

- A lognormal distribution satisfactorily describes the distribution of maintenance times.
- The variance of the logarithms of the maintenance times is already known.

### 10.5.5 Method V: The man-hour rate (using simulated faults) method

This method demonstrates man-hour rates or man-hours per operating hour and is based on the following three items:

- **Item 1:** Predicted equipment failure rate.
- **Item 2:** Total accumulative chargeable maintenance man-hours.
- **Item 3:** Total accumulative simulated demonstration operating hours.

### 10.5.6 Method IV: The man-hour method

This test method demonstrates man-hour rates, specifically per flight hour. It makes use of the accumulative flight hours and the determination, during Phase II test operation, of the total accumulative chargeable maintenance man-hours. It is to be noted that in using this method:

- Ensure that the predicted man-hour rate clearly pertains to flight time rather than the equipment operating time.
- Develop proper ratios of equipment operating time to flight time.

### 10.5.7 Method VII: The percentiles and maintenance method

This method makes use of a test of proportion for demonstrating fulfillment of maximum preventive maintenance task time at any percentile, median corrective maintenance task time, and median preventive maintenance task time, in situations when corrective and preventive repair time distributions are unknown. The following two factors are associated with the method:

- **Factor I:** A maximum sample size of 50 tasks is needed.
- **Factor II:** The plan holds the confidence level at 75% or 90%.

### 10.5.8 Method VIII: The preventive maintenance times method

This method is considered quite useful when the specified index involves mean preventive maintenance task time and/or maximum preventive maintenance task time at any percentile, and when all possible preventive maintenance tasks need to be accomplished. The test needs no allowance for assumed statistical distribution.

### 10.5.9 Method IX: The chargeable maintenance downtime per flight method

This method uses the Central Limit Theorem and is used in testing aircraft. The chargeable downtime per flight is the allowable time, stated in hours, for conducting maintenance assuming that there is a specific availability and operation readiness requirement for the aircraft.

### 10.5.10 Method X: The combined mean/percentile requirement method

This method is considered quite useful when the specification is expressed as a dual requirement for the mean and for either the 90th or 95th percentile of maintenance times, when maintenance time is lognormally distributed.

### 10.5.11 Method XI: The median equipment repair time method

Maintainability demonstrations make use of this method when the need or requirement is expressed in terms of an equipment repair time median. The method is based on log normally distributed corrective maintenance times as well as a sample size of 20.

## 10.6 Steps for performing maintainability demonstrations and evaluating the results and guidelines to avoid pitfalls in maintainability testing

There are many steps associated with the preparation for conducting maintainability demonstrations and evaluating their results. All these steps are described below [2, 7].

- **Step 1:** This step is concerned with selecting the test method or methods outlined in MIL-STD-471A [4]. This selection depends on factors such as product characteristics and the parameters to be demonstrated.

- **Step 2:** This step is concerned with establishing accept/reject criteria and retest procedures in the event that the 'accept' criteria in not met.
- **Step 3:** This step is concerned with developing in detail a maintainability demonstration plan and test procedures. The plan addresses issues such as equipment and documentation required, facility needs, manpower requirements, and training needs.
- **Step 4:** This step is concerned with choosing maintenance task population out of which the maintainability test sample will be taken.
- **Step 5:** This step is concerned with the pretest preparation. This includes preparing the facilities needed for the test and assembling the test hardware, test support equipment, documentation, and other requirements.
- **Step 6:** This step is concerned with performing the maintainability demonstration test or tests.
- **Step 7:** This step is concerned with performing post-test tasks such as restoring test hardware to its original form, verifying the test hardware's acceptability for use on production items, and returning test equipment and associated facilities to pretest form.
- **Step 8:** This step is concerned with the analysis of the test data, which includes determining whether acceptance criteria were satisfied and analyzing the maintenance strengths and weaknesses of the product.
- **Step 9:** This step is concerned with recommending corrective measures as appropriate.
- **Step 10:** This step is concerned with preparing the documentation related to the demonstration test.

Various studies conducted over the years indicate that there are various types of pitfalls that can occur in maintainability testing process. Some of the useful guidelines to avoid pitfalls in maintainability testing are as follows [2, 8]:

- Conduct some 'dry run' testing, if possible.
- Improve the technical manual verification and validation process before the maintainability demonstration test.
- Tailor MIL-STD-471 [4] for the program and product/system under consideration rather than totally relying on it.
- Perform a new and different trial for each and every trial that highlights a deficiency.
- Properly define, verify, and correct all discovered deficiencies and the associated needs for corrective action.
- Limit the allowable trial repetitions as a requirement to cancel the test, progressing into an 'evaluate and fix' phase, and then repeating the test with newly stated faults.

## 10.7 Problems

1. List the six categories of maintainability testing and demonstration planning and control requirements. Describe three categories in detail.
2. What are the items that maintainability demonstrations plans and procedures should cover?
3. What are items related to test results that the maintainability demonstration reports include?
4. Describe the following two maintainability test approaches:
   - Dynamic tests.
   - Static tests.
5. Describe the steps in preparing for and conducting a maintainability demonstration and in evaluating its results.
6. Describe the following two maintainability testing methods:
   - The mean method.
   - The critical percentile method.
7. Discuss the useful guidelines to avoid pitfalls in maintainability testing.
8. Describe the following maintainability test approaches:
   - Closed-loop tests.
   - Open-loop tests.
   - Marginal tests.
9. List the maintainability testing methods described in Ref. [4] (i.e., MIL-STD-471).
10. Describe the following two maintainability testing methods:
    - The critical maintenance time or man-hours method.
    - The mean maintenance time and maximum maintenance time method.

## References

1. MIL-STD-470, Maintainability Program Requirements, Department of Defense, Washington, D.C., 1966.
2. Dhillon, B.S., Engineering Maintainability, Gulf Publishing, Houston, Texas, 1999.
3. AMCP-706-133, Maintainability Engineering Theory and Practice, Department of Defense, Washington, D.C., 1976.
4. MIL-STD-471, Maintainability Verification/Demonstration/Evaluation, Department of Defense, Washington, D.C., 1966 (revision "A", 1973).
5. RADC Reliability Engineer's Toolkit, Systems Reliability and Engineering Division, Rome Air Development Center (RADC), Griffiss Air Force Base, New York, 1988.

6. PRIM-1, A Primer for DOD Reliability, Maintainability and Safety Standards, Reliability Analysis Center, Rome Air Development Center, Griffiss Air Force Base, New York, 1988.
7. Grant Ireson, W., Coombs, C.F., Moss, R.Y., Handbook of Reliability Engineering and Management, McGraw-Hill, New York, 1996.
8. Bentz, R.W., Pitfalls to Avoid in Maintainability Testing, Proceedings of the Reliability and Maintainability Symposium, 1982, pp. 278–282.

# *chapter eleven*

# *Safety management*

## *11.1 Introduction*

Safety is an acknowledged management responsibility, and each person has responsibility for his or her own safety in addition to that of others whom his or her actions may affect directly or indirectly. In situations where work is being accomplished through the organization of personnel/people, the safety of these personnel becomes the responsibility of the line of authority or management. Thus, the safety management's fundamental objective is eradicating human anguish and suffering and achieving economy of operations in an effective manner.

The beginning of the real safety management may be regarded as the period during the 1950s and 1960s [1]. An important milestone in the history of safety management was the passage of the Occupational Safety and Health Act (OSHA) by the U.S. Congress in 1970. Since 1970, many new developments related to safety management have taken place, and safety management may simply be defined as the accomplishment of safety through the efforts of others [2–4].

This chapter presents various important aspects of safety management.

## *11.2 Principles of safety management*

Over the years, professionals working in the area of safety have developed various principles of safety management. Some of these principles are as follows [5–7]:

- The key to successful line safety performance is management procedures that effectively factor in accountability.
- The symptoms that indicate something is wrong in the management system are an unsafe condition, an accident, and an unsafe act.
- Safety should be managed just like any other function in an organization. More clearly, management should direct safety by setting attainable goals, by planning, organizing, and controlling for attaining them successfully.

- There is no single approach for effectively achieving safety in an organization. But, for a safety system to be effective, it must effectively satisfy certain criteria: involves worker participation, be flexible, and have the top management visibly showing its full support; hence, supervisory performance involves middle level management personnel and must be perceived as positive.
- The function of safety is discovering and defining the operational errors that lead to accidents.
- Under most circumstances, unsafe behavior is general human behavior because it is the result of normal people reacting to their surrounding environment. Therefore, it is the management's responsibility for making changes to the environment that leads to the unsafe behavior.
- There are specific sets of circumstances that can be predicted to result in severe injuries: certain construction-related situations, non-productive activities, high energy sources, and unusual, non-routine activities.
- The causes that lead to unsafe behavior can be highlighted, controlled, and classified. Some classifications of the causes include overload, the employee's decision to err, and traps.
- In building an effective safety system, the three major subsystems that must be properly considered are the behavioral, the physical, and the managerial.
- The safety system should be made to fit properly to the organization/company culture.

## 11.3 Functions of safety department, manager, and engineer

Although the functions of a safety department may vary from company/organization to another, however, some of its typical functions are as follows [3, 7, 8]:

- Develop and administer the organization/company safety program.
- Prepare reports concerning company/organization safety performance and justify safety-related measures.
- Provide safety training.
- Liaise with others on safety-related matters (e.g., insurance companies and governmental agencies).
- Evaluate company compliance with governmental and other regulations in regard to safety.
- Procure and distribute personal protective equipment.
- Carry out safety inspections and surveys.
- Investigate accidents.

- Publicize safety-related materials.
- Keep data on work-related injuries and illnesses.

There are many functions of a safety manager, and they may vary from one company to another. However, some of the important functions of a safety manager are as follows [3, 7, 9]:

- Formulating and administering the safety program.
- Reporting to upper level management periodically, concerning the state of the company safety-related effort.
- Advertising safety-related issues at all levels of management.
- Participating in the review of specifications as well as in the design of new facilities/equipment layout/process layout.
- Directing the inspection of the facilities for proper compliance with the rules and regulations of the outside bodies.
- Representing management to public, government agencies, insurance companies, employees, etc., in regard to safety.
- Supervising all employees in the safety department.
- Directing the collection and recording of important information on matters such as work-related injuries and accidents.
- Acquiring the most up-to-date and best hazard control-related information.

There are various types of functions of a safety engineer and they may vary from one company to another and even within a company. Nonetheless, some of the functions of a typical safety engineer are as follows [7, 8]:

- Perform accident-related investigations.
- Conduct safety inspections.
- Collaborate with safety committees.
- Carry out safety-related studies.
- Coordinate with management on matters concerning safety.
- Ensure that the appropriate measures are taken for avoiding accidents.
- Provide safety training.
- Process workers' compensation claims.

## 11.4 *Steps for developing a safety program plan and managerial-related deficiencies leading to accidents*

An organization contemplating on introducing a safety program can develop its plan by following the seven steps shown in Fig. 11.1 [3, 7].

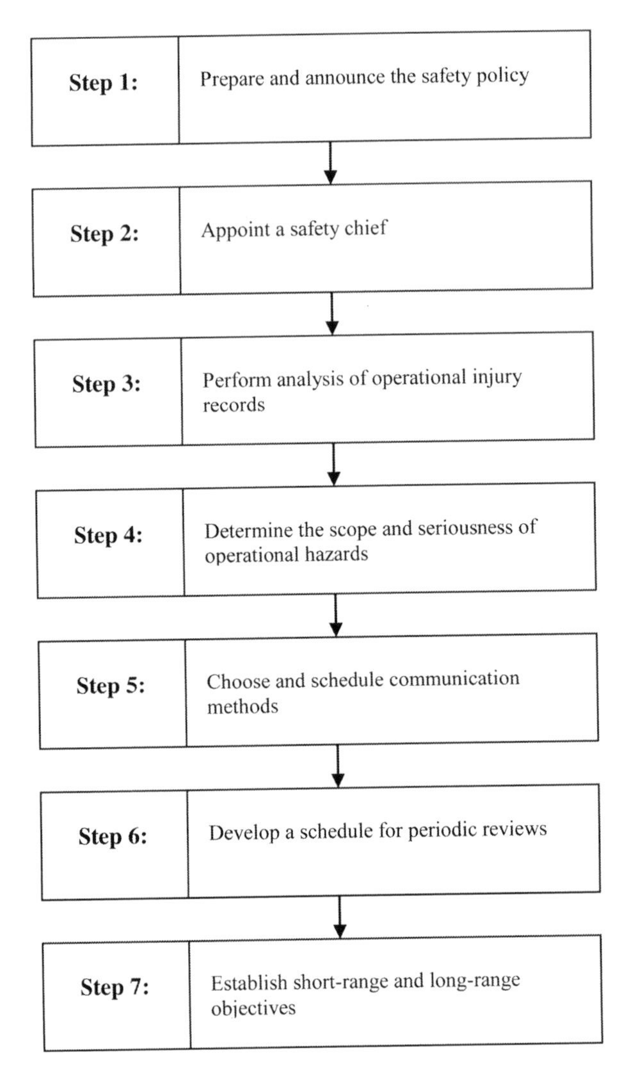


***Figure 11.1*** Steps for developing a safety program plan.

Step 1 is concerned with writing and announcing the policy regarding the control of hazards within the organization/company and designating accountability and authority for fully implementing it. Step 2 is concerned with appointing a safety chief for looking after safety-related matters. Step 3 is concerned with conducting analysis of the operational records of injuries, property damage, and work-related illnesses.

Step 4 is concerned with determining the scope and seriousness of operational hazards. More clearly, the step is concerned with determining the quality of the existing physical safeguards, time estimates and budgets for conducting the corrective measures, the nature and severity of inherent operating hazards, the required corrective measures, etc.

Step 5 is concerned with selecting and scheduling communication methods for purposes such as informing general management the safety progress of the organization and associated requirements, worker safety training, and interest maintenance. Step 6 is concerned with developing a schedule for periodic reviews of the facilities and program. Finally, step 7 is concerned with developing short-range and long-range objectives for the safety program.

Past experiences over the years clearly indicate that many accidents take place due to various types of management deficiencies. On the other hand, as per Fortune magazine, many executives still believe that careless workers/employees are really to be blamed for the workplace accidents' occurrence [10]. However, the actual findings do not support the beliefs of these executives. For example, a survey conducted in 1967 of industrial injuries in the state of Pennsylvania reported that only about 26% were due to workers' carelessness [11].

Eight immediate causes for various accidents due to management-related deficiencies are as follows [12]:

- **Immediate accident cause I:** Improper use of equipment, tools, and facilities. In this case, possible management deficiencies are poor training of workers and poorly established operational procedures.
- **Immediate accident cause II:** Job not understood. In this case, possible management deficiencies are poorly written operational procedures and poor employee selection and placement.
- **Immediate accident cause III:** Lack of awareness of hazards involved. In this case, possible management deficiencies are poor worker training, poor worker safety consciousness, and poor safety rules and measures.
- **Immediate accident cause IV:** Failure to follow prescribed procedures correctly. In this case, possible management deficiencies are poor enforcement to follow correct procedures and poor supervisory safety indoctrination.
- **Immediate accident cause V:** Defective or unsafe facilities and equipment. In this case, possible management deficiencies are poor supervisory safety indoctrination, poor employee safety indoctrination, and poor maintenance and repair system.
- **Immediate accident cause VI:** Lack of proper procedures. In this case, possible management deficiencies are poor supervisory proficiency; poor planning, layout, and design; and poor operational procedures.
- **Immediate accident cause VII:** Lack of proper equipment, tools, and facilities. In this case, possible management deficiencies are poor supervisory safety indoctrination and poor planning, layout, and design.

- **Immediate accident cause VIII:** Poor housekeeping. In this case, possible management deficiencies are poor supervisory training and poor planning and layout.

## 11.5 Product safety management program and organization tasks

The main objective of a product safety management (PSM) program is to minimize an organization's/company's exposure to product liability litigation and related problems. Thus, the key for minimizing liability exposure is to establish and properly maintain a comprehensive safety management program. Generally, a PSM program has at least three functional elements, as shown in Fig. 11.2 [2, 7]. These elements are PSM program committee, coordinator, and auditor. The committee is formed for dealing with issues concerned with product safety.

As the committee has members from all major units within the organization/company, it provides the PSM coordinator a broad base of expertise to call upon and encourages broad-based support among all involved units.

The coordinator is concerned with coordinating and facilitating the involvement of various units within the company/organization including the marketing, design, manufacturing, accounting, and service. In the success or failure of a PSM program, the level of authority of the coordinator plays a crucial factor. Past experiences over the years clearly indicate that the higher the level of authority, the greater the chances the program will succeed.

The National Safety Council (NSC) recommends that the PSM coordinator should have authority for undertaking actions such as assisting in developing PSM program-related policy; performing PSM program audits; coordinating all program-related documents; making recommendations for product recalls, special analysis, product redesign, and field modifications; facilitating communication among all parties

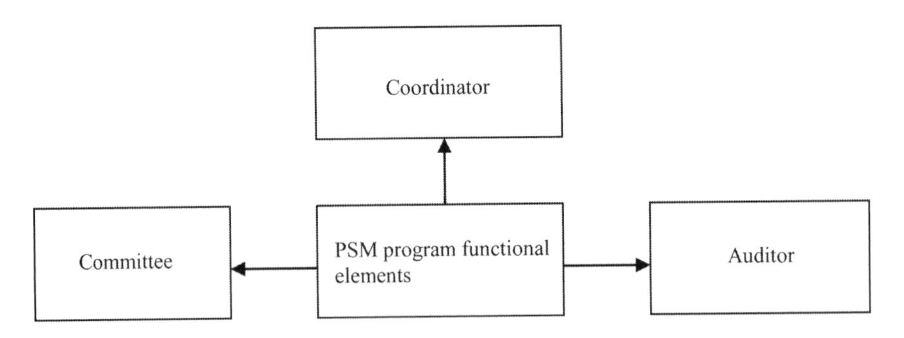


*Figure 11.2* Functional elements of a PSM program.

involved with the program; performing complaint, incident, or accident analysis; establishing and maintaining relationships with agencies/organizations having product safety-related missions; and developing a product safety information database for use by all parties involved in the program [2, 7, 13].

Finally, the auditor is concerned with evaluating the overall organization and units within it in regard to safety. Nonetheless, the auditor's specific duties include highlighting deficiencies in management commitment, reviewing documentation of measures taken for rectifying product shortcomings, bringing deficiencies to management attention and making appropriate corresponding recommendations, and observing the corrective measures put in place after the identification of product deficiencies [2, 7].

A product safety organization conducts various types of tasks concerning product safety. Some of these tasks are as follows [7, 14]:

- Review governmental and non-governmental product safety-related requirements.
- Develop safety criteria based on applicable governmental and voluntary standards for use by company, sub-contractor, and vendor design professionals.
- Prepare the product safety directives and the program.
- Review product test reports for determining deficiencies or trends in regard to safety.
- Review warning labels that are to be placed on product in regard to safety factors such as meeting legal requirements, adequacy, and compatibility to warnings in the instruction manuals.
- Take part in reviewing accident-related claims or recall actions by government bodies and recommend remedial measures for justifiable claims or recalls.
- Review the product for determining whether the potential hazards have been controlled or eradicated altogether.
- Provide assistance to designers in choosing alternative means for controlling or eradicating hazards or other safety problems in initial designs.
- Develop mechanisms by which the safety program can be monitored properly.
- Review hazard and mishaps in existing similar products for avoiding their repetition in the new products.
- Review safety-related customer complaints and field reports.
- Review proposed product operations and maintenance documents in regard to safety.
- Determine whether items such as monitoring and warning devices, protective equipment, or emergency equipment are required for the product.

## 11.6 Safety performance measures and drawbacks of the standard indexes

Management uses various types of indexes for measuring the organization's safety performance. This section presents two such indexes taken from a standard developed by the American National Standards Institute in 1985 [15].

### 11.6.1 Index I: Disabling injury severity rate

This is a commonly used index and is defined by [15]

$$DISR = \frac{DC_t(1{,}000{,}000)}{EET} \tag{11.1}$$

where

$DISR$ is the disabling injury severity rate.
$DC_t$ is the total days charged.
$EET$ is the employee exposure time expressed in hours.

It is to be noted that this index is based on four factors occurring during the period covered by the rate. These factors are the total scheduled charges (days) for all deaths, total scheduled charges (days) for all permanent disabilities, total scheduled charges (days) for all permanent partial disabilities, and the total days of disability from all temporary injuries. Some of the advantages of the index are as follows:

- It can be used for making comparisons among different organizations.
- It takes into consideration differences in quantity of exposure over time.
- It clearly answers the question 'How serious are injuries in our organization/company'.

### 11.6.2 Disabling injury frequency rate

This is expressed by [15]

$$DIFR = \frac{DI_t(1{,}000{,}000)}{EET} \tag{11.2}$$

where

$DIFR$ is the disabling injury frequency rate (DIFR).
$DI_t$ is the total number of disabling injuries.
$EET$ is the employee exposure time expressed in hours.

It is to be noted that this index is based on four events that take place during the time period covered by the rate: temporary disabilities (total), permanent partial disabilities (total), permanent disabilities (total), and deaths (total). The main advantage of this index is that it takes into consideration differences in quantity of exposures due to varying employee/worker work-related hours, either within the framework of the organization during successive periods or among organizations within similar industry categories [16].

### 11.6.3 *Drawbacks of the standard indexes*

There are many drawbacks of the standard indexes. Some of these drawbacks are as follows [15, 16]:

- Deaths, recordable occupational-related illnesses and injuries, lost-time accidents, and other injuries reported are relatively quite rare events. More clearly, small organizations/companies/units may not record a reportable accident or incident under the existing system of measurement for long period.
- Many accidents, in particular the less severe ones, are never reported; thus, the indexes exclude this quite valuable information.
- Insufficient sensitivity to be employed as accurate indicators of safety effectiveness due to inclusion of only those illnesses or accidents in computations that result in actual losses.
- For smaller work forces, the indexes' reliability decreases quite significantly.

## 11.7 *Problems*

1. List at least ten principles of safety management.
2. What are the functions of a safety manager?
3. Define the term 'safety management', and write an essay on safety management.
4. Discuss the steps involved in developing a safety program plan for an organization.
5. What are the important functions of a safety engineer?
6. Discuss at least eight immediate causes for accidents due to management deficiencies.
7. What are the functional elements of a PSM program? Discuss each of these elements in detail.
8. What are the important tasks of a product safety organization?
9. Define and discuss the following safety-related index:
   - DIFR.
10. What are the drawbacks of the standard indexes used for measuring the safety performance of an organization?

## References

1. Petersen, D., Safety Management: A Human Approach, Aloray Publisher, Englewood Cliffs, New Jersey, 1975.
2. Goetsch, D.L., Occupational Safety and Health, Prentice Hall, Englewood Cliffs, New Jersey, 1996.
3. Grimaldi, J.V., Simonds, R.H., Safety Management, Richard D. Irwin, Chicago, 1989.
4. Roland, H.E., Molriarty, B., System Safety Engineering and Management, John Wiley and Sons, New York, 1983.
5. Petersen, D., Safety Management, American Society of Safety Engineers, Des Plaines, Illinois, 1998.
6. Petersen, D., Techniques of Safety Management, McGraw-Hill, New York, 1971.
7. Dhillon, B.S., Engineering Safety: Fundamentals, Techniques, and Applications, World Scientific Publishing, River Edge, New Jersey, 2003.
8. Gloss, D.S., Wardle, M.G., Introduction to Safety Engineering, John Wiley and Sons, New York, 1984.
9. Schenkelbach, L., The Safety Management Primer, Dow Jones-Irwin, Homewood, Illinois, 1975.
10. Cordtz, D., Safety on the Job Becomes a Major Job for Management, Fortune, November 1972, p. 42.
11. Hammer, W., Price, D., Occupational Safety Management and Engineering, Prentice Hall, Upper Saddle River, New Jersey, 2000.
12. Peters, T., Thriving on Chaos: Handbook for a Management Revolution, Harper and Row, New York, 1987.
13. Accident Prevention Manual for Industrial Operations, National Safety Council (NSC), Chicago, 1988.
14. Hammer, W., Product Safety Management and Engineering, Prentice Hall, Englewood Cliffs, New Jersey, 1980.
15. Z.16.1, Methods of Recording and Measuring Work Injury Experience, American National Standards Institute, New York, 1985.
16. Tarrants, W.E., The Measurement of Safety Performance, Garland STPM Press, New York, 1980.

# chapter twelve

# Safety costing

## 12.1 Introduction

The history of safety costing may be traced back to 1944, when R.B. Blake, a safety engineer from the U.S. Department of Labor pointed out that 'the major driving force behind the industrial safety movement is the fact that accidents are costly' [1, 2].

As per Ref. [3], in 2000 in the United States, there were a total of 97,300 unintentional injury deaths and a disabling injury occurring every 1.5 seconds, and their total cost was estimated to be around $512.4 billion, in turn translating to about $5,000 per household. This cost figure includes items such as medical-related expenses, property damage, fire losses, and employer costs.

All in all, it may be said that the rising cost of accidents is a very important factor in the increasing attention on safety. This chapter presents various important aspects of safety costing.

## 12.2 Safety cost-related facts, figures, and examples

Some of the facts, figures, and examples directly or indirectly concerned with the safety cost are as follows:

- In 1995, workplace accidents alone cost the United States approximately $75 billion [4].
- In 1980, the U.S. Department of Health and Human Services reported that employers spent approximately $22 billion to insure or self-insure against job-associated injuries [5].
- In 1981, there were approximately 99,000 accidental deaths and 9,400,000 disabling injuries in the United States. Their combined cost was estimated to be about $87.4 billon [6].
- In 2000, there were a total of 97,300 unintentional injury deaths and a disabling injury occurring every 1.5 seconds in the entire United States. Their overall total cost was estimated to be about $512.4 billion [3].

- In 1996, due to a fuel-tank fire, a Paris-bound Trans World Airlines jet crashed and killed all 230 persons onboard. A subsequent task force concluded that by adding non-flammable gases (fuel-tank inverting) would significantly decrease the risk of fuel-tank explosions but recommended against such changes because of the enormous cost tag. The task force estimated the cost of fuel-tank modification would be somewhere between $10 billion and $20 billion, and only a total of 253 lives would be saved [7].
- In 1996, the National Safety Council (NSC) stated that the value of goods or services each worker/employee must produce for offsetting the cost of work-related injuries is around $960 [8].
- In 2000, the total estimated cost of work-related injuries to the entire United States was around $131.2 billion. This figure exceeds the combined profits of the top 13 Fortune 500 companies [3].
- In 1995, the Occupational Safety and Health Administration (OSHA) issued initial penalties of at least $100,000 against 122 employers [8].
- As per [8] in 1970, an injury made the highest award of $3,650,000 in legal history to date received by a young diesel mechanic when bolt snapped and a falling piece of equipment crushed part of his skull [8].
- In 1983, the capital expenditure investment of American businesses for employee safety and health was estimated to be around $5.7 billion, and it was predicted to grow by 10% for the next number of years [6, 9].
- In 1997, three workers were awarded almost $5.8 million after they sued a computer equipment manufacturer for musculoskeletal disorders (MSDs) because they believed that these disorders were due to keyboard entry activities [8].
- As per [8], the direct or indirect cost of an accident in 1979, at the Three Mile Island nuclear power plant, was estimated to be approximately $4 billion.
- In 1993, a Virginia jury awarded $8 million to a worker for back-related injuries he received when a piece of equipment fell onto him [8].

## 12.3 Losses of a company due to an accident involving its product

There are a large number of losses to a company that can result from accidents involving its manufactured products. Most of these losses are as follows [10, 11]:

- Cost of corrective measures necessary for preventing reoccurrences.
- Accident investigation cost.

- Cost of slowdowns in operations while accident-related causes are being investigated and remedial measures taken.
- Payments for property damage-related claims that are not covered by insurance. These claims may include items such as administrative-related cost, replacement costs for the product, cost of emergency assistance, lost time and wages, loss of operation time, and plaintiff's legal fee.
- Payments for death or injury claim settlements.
- Increase in insurance cost.
- Loss of prestige.
- Cost of legal aid for defense against claims.
- Involved manufacturer's personnel lost time.
- Penalties for failure to take proper measures for rectifying hazards/defects/conditions violating statues.
- Loss of revenue because of degradation in public confidence.
- Degradation in morale.
- Punitive damages.

Finally, it is to be noted that all the above losses may or may not be applicable to a given product.

## 12.4 Safety cost estimation methods

Over the years, several methods have been developed for estimating various types of safety costs. Three of these methods are presented below.

### 12.4.1 The Simonds method

This method was developed by Professor R. H. Simonds of Michigan State College when working in conjunction with the NSC [12, 13]. Simonds reasoned that an accident's cost can be grouped under two main categories: insured and uninsured costs. Furthermore, he pointed out that the insured cost can easily be estimated by simply examining some accounting records, but the estimation of the uninsured costs is quite challenging. Thus, Simonds proposed the following equation for estimating the total uninsured cost of accidents [11–13]:

$$TUC = \alpha_1 AC_1 + \alpha_2 AC_2 + \alpha_3 AC_3 + \alpha_4 AC_4 \quad (12.1)$$

where

TUC is the total uninsured cost of accidents.

$\alpha_1$ is the total number of lost work-day cases due to Class 1 accidents resulting in permanent partial disabilities and temporary total disabilities.

$AC_1$ is the average uninsured cost associated with Class 1 accidents.

$\alpha_2$ is the total number of physician's cases associated with Class 2 accidents, i.e., the Occupational Safety and Health Act (OSHA) non-lost work-day cases requiring treatment by a doctor.
$AC_2$ is the average uninsured cost associated with Class 2 accidents.
$\alpha_3$ is the total number of first-aid cases associated with Class 3 accidents, i.e., those accidents in which first-aid was provided locally with a loss of working time of less than 8 hours.
$AC_3$ is the average uninsured cost associated with Class 3 accidents.
$\alpha_4$ is the total number of non-injury cases associated with Class 4 accidents, i.e., those accidents causing minor injuries that do not require the attention of a medical professional.
$AC_4$ is the average uninsured cost associated with Class 4 accidents.

### 12.4.2 The Heinrich method

Over 80 years earlier, H.W. Heinrich categorically stated that for each and every dollar of insured cost paid for accidents, there were $4 of uninsured costs borne by the company/organization [14]. His conclusions were based on factors such as the review of 5000 case files from organizations insured with a private company, interviews with the staff members of the administrative and production services of those enterprises, and research in the concerned organizations [15]. Heinrich expressed 'total cost of occupational-related injuries' as the sum of the direct and indirect costs. The direct cost is made up of the total benefits paid by the insurance company, and the indirect cost, the expenditure assumed directly by the enterprise and is made up of the following items [15]:

- Lost time cost of workers who stop work are involved in the action.
- Injured worker's lost time cost.
- Cost of overheads for injured worker while in non-production mode.
- Cost to workers under welfare and benefit system.
- Management's lost time cost.
- Lost time cost of first aid and hospital workers not paid by insurance.
- Cost to workers in continuing wages of the insured.
- Cost due to weakened morale.
- Material/Machine damage-related cost.
- Lost orders' cost.
- Cost associated with profit and worker productivity loss.

### 12.4.3 The Wallach method

This method was developed in 1962 by M.B. Wallach for analyzing the cost of the consequences of events concerning occupational injuries in various areas related to production: machines, equipment, man-power, materials, and time [16]. Although this method highlights only the occupational-related injuries' effect on production, but it has a distinct advantage of employing ideas and language familiar to companies/organizations.

All in all, the Wallach method appears to be quite effective, particularly when measures are taken for quantifying occupational injuries' effect at the company level [15]. Additional information on this method is available in [16].

## 12.5 Safety cost estimation models

Over the years, professionals working in the area have developed various general and specific safety cost estimation models. Four of these models are presented below.

### 12.5.1 Accident hidden cost estimation model

There are many hidden costs associated with accidents. The total hidden cost of an accident is defined by [11]

$$THC_a = C_d + C_{wu} + C_{wi} + C_{eo} + C_s + CL_n + C_u + C_w + C_h + C_m \qquad (12.2)$$

where

$THC_a$ is the total hidden cost of an accident.
$C_d$ is the cost of damage to equipment/material.
$C_{wu}$ is the cost of wages paid for time lost by uninjured workers.
$C_{wi}$ is the cost of wages paid for time lost by the injured workers.
$C_{eo}$ is the extra cost of overtime work necessitated by the occurrence of the accident.
$C_s$ is the cost of wages paid to supervisory personnel for time needed for activities necessitated by the occurrence of the accident.
$CL_n$ is the cost of the learning period of the new worker replacing the injured one.
$C_u$ is the uninsured medical cost borne by the company/enterprise.
$C_w$ is the wage cost due to reduced output of an injured worker after his/her return to work.
$C_h$ is the cost of time spent by higher supervisory and clerical personnel.
$C_m$ is the miscellaneous general cost.

It is to be noted that the right-hand elements of Equation (12.2) may vary quite significantly from one accident to another. More clearly, the values of some of these elements could be equal to zero.

### 12.5.2 *Total safety cost estimation model*

This model can be used for estimating total safety cost and it is expressed by [8, 11]:

$$SC_t = C_i + C_{ap} + C_{sl} + C_r + C_w + C_{im} + C_{in} + C_m \tag{12.3}$$

where
- $SC_t$ is the total safety cost.
- $C_i$ is the cost associated with immediate losses due to accidents.
- $C_{ap}$ is the cost associated with accident prevention-related measures.
- $C_{sl}$ is the cost associated with safety-related legal issues.
- $C_r$ is the cost associated with rehabilitation and restoration.
- $C_w$ is the cost associated with welfare-related issues.
- $C_{im}$ is the cost associated with the immeasurable.
- $C_{in}$ is the cost associated with insurance.
- $C_m$ is the cost associated with miscellaneous safety-related issues.

### 12.5.3 *Product life cycle safety cost estimation model I*

The safety cost of a product/system over its entire life cycle is defined by [10]

$$LCSC_p = APPC + RC + PC + IC - R \tag{12.4}$$

where
- $LCSC_p$ is the product/system lie cycle safety cost.
- $APPC$ is the accident prevention program cost.
- $RC$ is the recall cost.
- $PC$ is the program cost.
- $IC$ is the insurance cost.
- $R$ is the reimbursements.

### 12.5.4 *Product life cycle safety cost estimation model II*

This is another mathematical model that can also be used for estimating the safety cost of a product/system over its entire lifespan. In this case, the product/system life cycle safety cost is expressed by [11]

$$LCSC_p = SCRDP + SCPCP + SCOSP + SCRD_p \tag{12.5}$$

where
- $LCSC_p$ is the product/system life cycle safety cost.

*SCRDP* is safety-related cost during the research and development phase of the product/system. This cost is basically concerned with various studies carried out and actions taken in regard to safety during the research and development phase.
*SCPCP* is the safety-related cost during the production and construction phase of the product/system. This cost includes expenditure on safety-related measures or any other safety-related expense during the phase.
*SCOSP* is the safety-related cost during the operation and support phase of the product/system. This cost includes all the safety-related costs that occur during the phase.
$SCRD_p$ is the safety-related cost during the retirement and disposal phase of the product/system. This cost includes all amounts spent in regard to safety for disposing the product/system.

## 12.6 Safety cost performance measurement indexes

Over the years, many indexes have been developed for measuring the overall safety cost performance of an organization. To the best of author's knowledge, there is no single index adequate for determining the overall safety cost effectiveness of an organization; several indexes in combination can be used to serve this very purpose.

The true purpose of using these indexes is to indicate trends, using the past safety cost performance as a point of reference, and to encourage all involved individuals to improve over the past performance. This section presents three, considered to be quite useful, safety performance cost-related indexes [11, 15, 17].

### 12.6.1 Average cost per injury index

This index is used for determining the average cost per occupational injury in an organization/company and is defined by

$$AC_{oi} = \frac{TC_{oi}}{TN_{oi}} \quad (12.6)$$

where

$AC_{oi}$ is the average cost per occupational injury.
$TC_{oi}$ is the total cost of occupational injuries.
$TN_{oi}$ is the total number of occupational injuries.

### 12.6.2 *Average injury cost per profit dollar index*

This index is used for determining the average cost of occupational-related injuries per profit dollar in organization/company and is defined by

$$ACOI_{pd} = \frac{TC_{oi}}{TP_d} \tag{12.7}$$

where
- $ACOI_{pd}$ is the average cost of occupational-related injuries per profit dollar.
- $TC_{oi}$ is the total cost of occupational-related injuries.
- $TP_d$ is the total profit in dollars.

### 12.6.3 *Average injury cost per unit turnover index*

This index is used for determining the average cost of occupational-related injuries per unit turnover in an organization/company and is defined by

$$ACOI_{ut} = \frac{TC_{oi}}{TN_{ut}} \tag{12.8}$$

where
- $ACOI_{ut}$ is the average cost of occupational-related injuries per unit turnover.
- $TC_{oi}$ is the total cost of occupational-related injuries.
- $TN_{ut}$ is the total number of units' turnover (i.e., unit quantity, tons, etc.).

## 12.7 *Problems*

1. Discuss at least seven important safety cost-related facts, figures, and examples.
2. Write an essay on safety costing.
3. List at least ten losses of a company that can result from accidents involving its manufactured products.
4. Discuss the Simonds safety cost estimation method.
5. Compare the Simonds and Heinrich safety cost estimation methods.
6. Write down the equation for the accident hidden cost estimation model.
7. Define the average cost per injury index.
8. Discuss the elements of the total safety cost estimation model.
9. Compare the two product life cycle safety cost estimation models presented in the chapter.
10. Describe the Wallach method.

# *References*

1. Safety Subjects, Bulletin No. 67, Bureau of Labor Standards, Department of Labor, U. S., Government Printing Office, Washington, D.C., 1955, p. 14.
2. Grimaldi, J.V., Simonds, R.H., Safety Management, Richard D. Irwin, Homewood, Illinois, 1989.
3. Report on Injuries in America in 2000, National Safety Council (NSC), Itasca, Illinois, 2001.
4. Spellman, F.R., Whiting, N.E., Safety Engineering: Principles and Practice, Government Institutes, Rockville, Maryland, 1999.
5. Lancianese, F., The Soaring Costs of Industrial Accidents, Occupational Hazards, August 1983, pp. 30–35.
6. Ferry, T.S., Safety Program Administration for Engineers and Managers, Charles C. Thomas Publisher, Springfield, Illinois, 1984.
7. Williams, W.E., Safety at All Costs, WorldNetDaily.Com, Cave Junction, Oregon, September 5, 2001, pp. 1–3.
8. Hammer, W., Price, D., Occupational Safety Management and Engineering, Prentice Hall, Upper Saddle River, New Jersey, 2001.
9. Tenth Annual McGraw-Hill Survey of Investment in Employee Safety and Health, McGraw-Hill Publications Department, McGraw-Hill, New York, June 1982.
10. Hammer, W., Product Safety Management and Engineering, Prentice Hall, Englewood Cliffs, New Jersey, 1980.
11. Dhillon, B.S., Engineering Safety: Fundamentals, Techniques, Applications, World Scientific Publishing, River Edge, New Jersey, 2003.
12. Simonds, R.M., Estimating Accident Cost in Industrial Plants, Practices Pamphlet No. III, National Safety Council, Chicago, 1950.
13. Goetsch, D.L., Occupational Safety and Health, Prentice Hall, Englewood Cliffs, New Jersey, 1996.
14. Heinrich, H.W., Industrial Accident Prevention, McGraw-Hill, New York, 1931.
15. Andreoni, D., The Cost of Occupational Accidents and Diseases, International Labor Office, Geneva, Switzerland, 1986.
16. Wallach, M.B., Accident Costs: A New Concept, The Journal of the American Society of Safety Engineers, Vol. 7, July 1962, pp. 25–26.
17. Blake, R.P., Industrial Safety, Prentice Hall, Englewood Cliffs, New Jersey, 1964.

## chapter thirteen

# Human factors in safety

## 13.1 Introduction

Although the histories of both human factors and safety may be traced back to the ancient times, but in the modern context, their serious starting points appear to be in 1898 and 1931, respectively. In 1898, Frederick W. Taylor carried out various studies for finding the most appropriate designs for shovels, and in 1931, H.W. Heinrich published the first book on industrial safety [1–3]. By 1945, human factors engineering was recognized as a specialized discipline, and in 1970, the U.S. Congress passed the Occupational Safety and Health Act (OSHA). Today, both human factors and safety are quite well recognized disciplines around the globe.

Human factors in safety has become a very important issue because each day in the United States alone about 9000 workers sustain disabling injuries on the job and 137 workers die from work-related illnesses [4]. In addition to various other measures, the modern safety practice is for providing and ensuring the factors listed below [5].

- Safeguards and designs that will stop the injuries' occurrence in the event when an error is made that causes an accident.
- Procedures that will substantially lower the chances of errors by operators and others.
- Designs that will totally eradicate the possibility of occurrence of errors and accidents.

This chapter presents various important aspects of human factors in safety.

## 13.2 Job stress

Past experiences over the years clearly indicate that job stress plays an important factor in safety. The basic cause of many accidents has been highlighted as job stress. The degree of stress's influence may vary significantly from one person to another due to factors such as person's genetically inherited physiologic responsiveness, basic personality, and education, acquired experience, and personality characteristics.

### 13.2.1 *Occupational stressor's classifications and workplace stress effects*

Occupational stressors may be classified under four categories, as shown in Fig. 13.1 [6]. The classifications shown in Fig. 13.1 are workload related, occupational frustration related, occupational change related, and miscellaneous. The workload-related stressors are concerned with workload problems (i.e., work overload or work underload). In the case of work overload, the task or job needs to exceed the individual's capability, and in the case of work underload, the work conducted by an individual fails to provide appropriate stimulation. Some examples of work underload are lack of need of any intellectual-related input, repetitive performance of the same task, and lack of opportunity for using knowledge and skills acquired by an individual.

The occupational frustration-related stressors are concerned with problems of occupational frustration, and they occur in situations where the assigned task/job inhibits the meeting of specified objective. The factors that form elements of occupational-related frustration include bureaucracy difficulties, lack of proper communication, and poor career

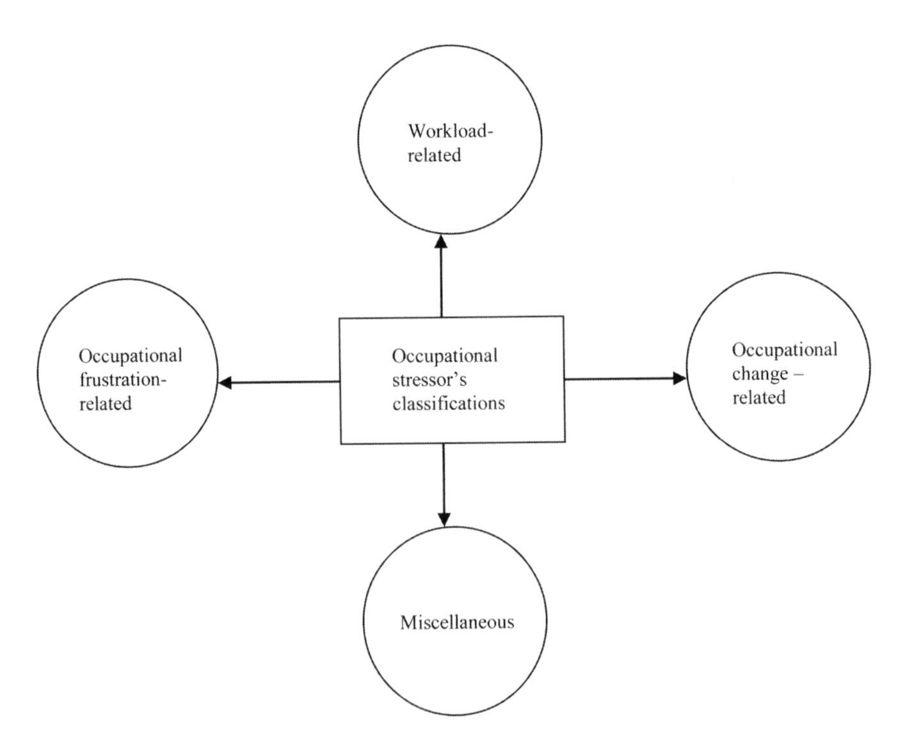


***Figure 13.1*** Types of occupational stressors.

development guidance. The occupational change-related stressors are concerned with an individual's behavioral, cognitive, and physiological patterns of functioning and are present in an organization/company concerned with productivity and growth. Some examples of occupational change are organizational restructuring, relocation, and promotion.

Finally, the miscellaneous stressors are stressors other than those belonging to the above three classifications. Some examples of these stressors are poor interpersonal relationships, too much or too little lighting, and noise.

There are many effects of workplace stress and they can be categorized under the following four classifications [3, 7, 8]:

- **Classification I:** This classification includes effects such as high turnover of workers, increased complainants and grievances, low morale translating into higher costs because of increased worker unrest, malicious destruction of equipment, etc.
- **Classification II:** This classification includes effects such as increased absenteeism, accidents, and alcoholism.
- **Classification III:** This classification includes effects such as increased rates of illness and disability in both worker and dependents leading to greater use of diagnostic and treatment services.
- **Classification IV:** This classification includes effects such as decreased production, quality deficiencies, and impaired decision making.

### 13.2.2 *Physical stress influencing factors*

There are essentially the following eight factors that influence stress [3, 9]:

- **Factor I: Sitting versus standing.** In this case, sitting in general is less stressful than standing. More clearly, standing for extended periods of time in one place can lead to unsafe levels of stress on the feet, legs, and back. Similarly, long sittings can also be quite stressful unless precautions such as proper posture, a supportive backrest, and frequent stretching/standing movements are properly taken.
- **Factor II: Low surface versus high surface.** In this case, past experiences clearly show that high surface contact-inclined tasks tend to be rather more stressful in a physical sense than low surface contact tasks. Surface stress can result from contact with hard surfaces such as tools and equipment.
- **Factor III: Large demand for strength/power versus small demand for strength/power.** In this case, it may be said that in general tasks that demand larger amounts of strength/power are rather more stressful than those requiring less.

- **Factor IV: Good vertical work area versus bad vertical work area.** In this case, a good vertical work area does not need the worker to bend forward or to twist the body from side to side but the bad one does. Consequently, the bad vertical work areas increase quite significantly the likelihood of physical stress.
- **Factor V: Stationary versus moveable/mobile jobs.** In this case, stationary jobs are basically performed at one workstation, and in contrast, mobile jobs need continual movement from one station to another. The potential for physical stress increases quite significantly with stationary tasks when workers/employees overlook to take precautions such as periodically stretching/standing/moving. Similarly, the potential for physical stress also increases significantly with mobile jobs when employees/workers carry items as they move from one station to another.
- **Factor VI: Good horizontal work area versus bad horizontal work area.** In this case, a good horizontal work area does not need the worker/employee to bend forward or to twist the body from side to side as opposed to the bad horizontal work area that needs such. In turn, it may be added that horizontal work surfaces increase significantly the potential for physical stress.
- **Factor VII: Non-repetitive motion versus repetitive motion.** In this case, it may be said that repetition can lead to monotony and boredom; thus, there is greater potential for physical-related stress.
- **Factor VIII: No negative environmental factors versus negative environmental factors.** In this case, it can be concluded that the more environmental-related factors that have to be contended with on the job, the more stressful the job.

Finally, it may be said that the higher the physical stress associated with a task, the greater the potential is for the occurrence of accidents.

### 13.2.3 *Human operator's stress-related characteristics and a checklist of stressors*

As a human operator possesses certain limitations in conducting a task, when those limitations exceed, the chances for making errors increase quite dramatically. In turn, the probability of accidents' occurrence also increases quite significantly. Nonetheless, some of the operator stress-related characteristics are as follows [10]:

- The required steps must be conducted at high speed.
- The operator's decision-making time is very short.
- The operator must make decisions on the basis of data obtained from various diverse sources.

- The operator is expected to make quick comparisons of more than one displays.
- The operator must operate two or more controls simultaneously at high speed.
- Inadequate information feedback to the operator for determining accuracy of his/her actions.
- The sequence of steps required is very long in conducting a task.
- Need for prolonged monitoring by the operator.

There are many stressors that can directly or indirectly influence a person's safe performance. A checklist of 12 of these stressors is as follows [3, 6]:

- **Stressor I:** Having inadequate expertise for performing the task being executed.
- **Stressor II:** Unhappiness with the current work or job.
- **Stressor III:** Currently working at a job that is well below one's professional qualifications.
- **Stressor IV:** Rumors of redundancy at work.
- **Stressor V:** Conducting tasks under time pressures.
- **Stressor VI:** Working with people having unpredictable temperaments.
- **Stressor VII:** Excessive demands from higher authority at the workplace.
- **Stressor VIII:** Regularly taking work home for meeting deadlines.
- **Stressor IX:** Limited scope for promotion at work.
- **Stressor X:** Poor health.
- **Stressor XI:** Passing through serious financial-related difficulties.
- **Stressor XII:** Experiencing difficulties with spouse or children or both.

## 13.3 *Work site analysis program for human factors*

The objective of the work site analysis program is to conduct work site human factors-related analysis and highlight stressors in the workplace. The program is composed of the following four main steps [11]:

- **Step I: Collect data.** This step is concerned with collecting information necessary for highlighting human factors-related hazards in workplace. Some of the sources for collecting such information are safety records, medical records, insurance records, and OSHA-200 logs.

- **Step II: Conduct baseline screening surveys.** These surveys help to highlight those jobs/tasks that put workers at risk of developing cumulative trauma disorders (CTDs). In situation when the job places workers at risk, there is absolutely a definite need for a program to require the human factors job hazard analysis.

  The survey is carried out with a human factors' checklist containing elements such as materials handling, posture, and upper extremity factors. The basis for the hazards' identification is human factor risk factors such as job process, conditions of a workstation, or work procedure that directly or indirectly contribute to the risk of developing CTDs. Some of the CTD risk factors are excessive vibration from power-tools, prolonged static postures, cold temperatures, continued physical contact with work surfaces, improper tools, repetitive or prolonged activities, awkward postures of the upper body, and forceful exertions.

  Similarly, the back disorder-related risk factors include poor grips on handles, prolonged sitting, bad body mechanics (e.g., continued bending over at the waist), lifting objects of excessive weights, and lack of adjustable chairs, body support, footrests, etc.
- **Step III: Conduct human factors job hazard analysis.** This is concerned with conducting human factors-related hazard analysis of each and every job/task that puts employees/workers at risk of developing CTDs. This type of analysis is considered extremely useful for verifying lower risk factors at light duty/restricted activity work positions as well as for determining whether risk factors for specific work position have been minimized or eliminated altogether.
- **Step IV: Conduct periodic surveys and follow-up studies.** These are basically concerned with conducting periodic reviews for highlighting factors/failures/deficiencies previously overlooked in work practices or engineering controls.

## 13.4 Symptoms of human factors-associated problems in organizations, identification of specific human factors-associated problems, and useful strategies for solving human factors-associated problems

Over the years, safety professionals have highlighted a number of symptoms of human factors-associated direct or indirect safety problems in organizations. Seven of these symptoms are as follows [11, 12]:

- **Symptom I: Cumulative trauma disorders (CTDs).** Some of the factors associated with CTDs are extreme temperatures, high levels of vibration and mechanical stress, a high level of repetitive work, greater than normal levels of hand force, and awkward posture. The degree of worker/employee exposure to factors such as these can be determined by observing the workplace as well as people/workers at work.
- **Symptom II: High absenteeism and turnover.** The high degree of absenteeism and turnover in an organization/company can be an indicator of human factors-associated problems. More clearly, it may be said that persons/individuals who are uncomfortable on the job to the point of experiencing physical-related stress are more likely to be absent from work and leave their current jobs for less stressful ones.
- **Symptom III: Work-created changes.** Past experiences over the years clearly indicate that workers tend to adapt to the workplace to satisfy their needs. The existence of large workplace adaptations, in particular the ones intended to lower physical stress, can indicate the existence of human factors-related problems. An example of these adaptations takes place when workers/employees have modified personal protective equipment and added padding.
- **Symptom IV: Visible trends in injuries and accidents.** By carefully examining items such as insurance forms, accident-related reports, OSHA-200 logs, and first aid logs safety professionals can establish the existence of trends. More clearly, a clear pattern or a high incidence rate of a certain type of injury normally is a good indicator of existence of human factors-associated problems.
- **Symptom V: Significant manual material handling.** Generally, the activities that involve a lot of manual material handling have higher incidence of musculoskeletal-related injuries. Furthermore, musculoskeletal-related injuries increase quite significantly in situations where factors such as lifting objects from the floor, bulky objects, lifting large objects, and high frequency of lifting are present. All in all, companies/organizations where factors such as these exist have human factors-associated problems.
- **Symptom VI: High incidence of worker complaints.** A high incidence of worker complaints concerning physical stress or improper workplace design can be a good indicator of human factors-associated problems.
- **Symptom VII: Poor quality.** This can result from human factors-associated problems, for example, poor inspection.

By conducting a task analysis of a given job, the specific human factors-associated problems can be identified. The problems that can be identified through task analysis include the following [12]:

- Tasks involving potentially hazardous movements.
- Tasks requiring unnatural or uncomfortable postures.
- Tasks with a high fatigue factor.
- Tasks with high potential for psychological stress.
- Tasks involving excessive wasted motion or energy.
- Tasks leading to quality control-related problems.
- Tasks due to a poor operations flow.
- Tasks involving frequent manual lifting.
- Tasks that should be automated.

Finally, it is added that some of the methods used for conducting task analysis are measuring the environment, questionnaires and interviews, drawing or sketching, photography and videotaping, and general observation [11].

Over the years, professionals working in the area of safety have developed various human factors problem-solving strategies for conducting tasks such as standing for heavy lifting/performing work in one place or in motion, seated repetitive work with light parts, seated work with larger parts, work with hands above chest height, standing work, seated control work, and work with hand tools. Thus, human factors problem-solving strategies for each of these seven tasks are as follows [9, 12, 13]:

- **Task I: Standing for heavy lifting/performing work in one place or in motion.** This type of task involves heavy lifting and moving materials in standing position and its associated common physical stress is muscle and back strains resulting from improper lifting. In this case, some of the human factors-associated strategies for improving work conditions include providing adequate room for facilitating lifting without twisting, minimizing manual carrying of heavy objects upstairs, minimizing manual lifting as much as possible by employing lifting and hoisting technologies, making proper personal protection equipment readily available, training workers/employees in the proper application of lifting methods, and keeping floor clean and dry where materials are to be lifted to prevent slips.
- **Task II: Seated repetitive work with light parts.** This type of task produces stress that generally leads to problems such as back, neck, lower leg, and shoulder pains. Some of the strategies for reducing these types of problems are to adjust work surface position or height; use ergonomics devices for adjusting the height and angle of work; incorporate other work tasks for breaking the monotony

of repetition; use an adjustable chair equipped with hand, wrist, or arm support as necessary; use job rotation with workers rotating from one or more different jobs; and ensure the existence of sufficient leg room in regard to depth, width, and height.

- **Task III: Seated work with larger parts.** Some of the problems associated with this type of task/work are related to areas such as posture, lifting, reach, and illumination. In this specific case, some of the human factors strategies for improving work-related conditions include using supplementary lighting at the workplace; using appropriate technology for lifting and positioning the work for easy access that does not need twisting, reaching, and bending; and using appropriate adjustable chairs and work surfaces.
- **Task IV: Work with hands above chest height.** Although this type of task/work can be conducted by either standing or sitting, its associated physical stress includes upper body, heart, and neck strain. Some of the strategies for reducing these types of problems are minimize manual lifting by raising the work floor using lifts and other technologies, use extension arms/poles when it is impossible to raise the work floor, and when procuring machines, look for machines with easily accessible controls below the horizontal plane of an employee's shoulders.
- **Task V: Standing work.** This type of task/work produces physical stress such as leg, back, and arm strains. Human factors-related strategies for improving work conditions include using adjustable machines and work surfaces for optimizing height and position, providing adequate space around machines for moving materials and ease of movement in servicing machines, and ensuring that in the procurement of new machines, there is a recess at the bottom for the feet.
- **Task VI: Seated control work.** This type of task/work involves using items such as levers, buttons, knobs, and wheels for controlling a system, process, or piece of equipment. In this case, some of the human factors-related strategies for improving work conditions are to use devices that fully satisfy standards such as finger control systems not requiring more than five newtons (1.1 pounds) and hand levers not exceeding 20 newtons (4.5 pounds), use an adjustable swivel chair with inflatable back and seat support, position the control seat in such a way that an absolutely clear line of sight exists between the back and the individual controlling it, and provide comfortable locations for control devices.
- **Task VII: Work with hand tools.** The usage of hand tools can introduce a variety of potential hazards, and the physical stress associated with the use of hand tools include muscle strains of the lower arm, wrist, and hand and carpal tunnel syndrome. Some of the human

factors-related strategies for improving the associated work conditions are to select tools with enhanced gripping surfaces on handles, minimize stress on the hand by selecting tools that have thick handles, choose tools specifically designed for keeping hands in the rest position, and select tools with handles having oval-shaped section when no twisting is involved.

## 13.5 Useful Occupational Safety and Health Administration (OSHA) ergonomics guidelines

The Occupational Safety and Health Administration (OSHA) agency has produced quite useful guidelines for general health and safety program management. Although these guidelines are voluntary, they provide quite useful information and guidance to companies for fulfilling their obligations under the U.S. Occupational Safety and Health Act concerning human factors. The agency particularly singled out the meat packing industrial sector for the human factors-related guidelines because of the high incidence of CTDs. Nonetheless, these OSHA guidelines have four major elements as shown in Fig. 13.2 [11, 14].

The main components of elements 'medical management' and 'education and training' are training for supervisors, training for managers, general training, job-specific training, and training for engineers and maintenance personnel. Similarly, four main components of elements

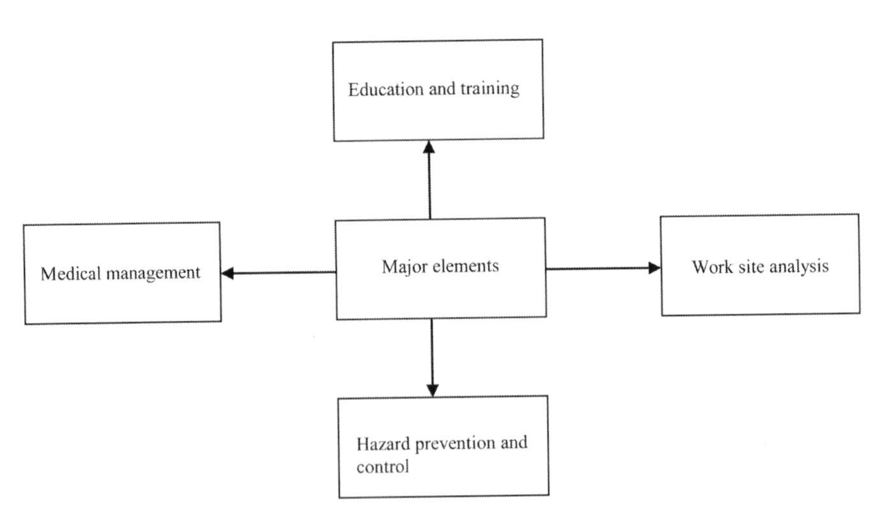


*Figure 13.2* Major elements of OSHA guidelines.

'work site analysis' and 'hazard prevention and control' are as follows [3, 11, 14]:

- Personal protective equipment.
- Work practice controls.
- Engineering controls.
- Administrative controls.

Finally, it is added that all the main components can be tailored for application in industries other than the meat packing industry [11].

## 13.6 *Human factors-related safety issues*

Over the years, professionals working in the safety area have pointed out various human factors-related issues that must be carefully considered early in a safety program because they may directly or indirectly affect safety. Ten of these issues are as follows [15]:

- **Issue I: Work space.** This issue is concerned with the appropriate adequacy of workspace for personnel and their tools and equipment.
- **Issue II: Displays and controls.** This issue is concerned with effective design and arrangement of controls and displays in regard to operators' and maintainers' natural sequence of operation-related actions.
- **Issue III: Environment.** This issue is concerned with an effective accommodation of environmental-related factors to which an item/system or human will be subjected.
- **Issue IV: Safety and health.** This issue is concerned with preventing operators', maintainers', and others' exposure to safety and health-related hazards.
- **Issue V: Training.** This issue is concerned with minimizing the need for maintainers', operators', and others' responses.
- **Issue VI: Special skills and tools.** This issue is concerned minimizing as much as possible the need for special or unique operator or maintainer skills, tools, abilities, or characteristics.
- **Issue VII: Communications.** This issue is concerned with effective system design-related considerations for enhancing required user communications and teamwork.
- **Issue VIII: Information requirements.** This issue is concerned with carefully considering the availability of information needed by individuals such as operators and maintainers for certain tasks when it is required and in the appropriate sequence.

- **Issue IX: Anthropometrics.** This issue is concerned with effective accommodation of individuals (i.e., from the 5th through 95th percentile levels of the human physical characteristics) represented in the user population by the system design.
- **Issue X: Oral/visual alerts.** This issue is concerned with effective designing of auditory and visual alerts for invoking operators', maintainers', and others' responses.

## 13.7 Employee training and education

The basic objective of safety training and education is for ensuring that employees are properly informed concerning the human factors-associated hazards to which they may be exposed. Safety-related training and education should start with the newly hired worker/employee and continue throughout his/her employment with the company/organization. Nonetheless, a safety-training program should include all affected individuals including workers, supervisors, and health care providers. The type of training, frequency, and materials presented may vary quite significantly from one organization/company to another. Nonetheless, safety training should include information on and review of items such as follows [5, 11]:

- Approaches for using emergency equipment.
- Safety-related rules, practices, and procedures developed by the company/organization.
- Means for preventing the occurrence of potential illnesses and injuries.
- Company/organization programs to aid workers/employees in procurement of safety-related personal equipment.
- Procedures for obtaining assistance in times of needs.
- Duties and rights of employees/workers under governmental safety-related legislations.
- Early symptoms and causes of potential illnesses and injuries.
- Need for strict observances of warning signs.
- Meanings of emergency-related signals.
- Illnesses' and injuries' potential risk.

## 13.8 Problems

1. What are the main categories of occupational stressors? Discuss each of these categories in detail.
2. Discuss the factors that influence physical stress.

3. What are the stressors that can directly or indirectly influence a person's safe performance?
4. What are the human operator's stress characteristics? List at least seven of them.
5. Discuss the main steps of the work site analysis program for human factors.
6. Discuss at least six symptoms of human factors-associated direct or indirect safety problems in organizations.
7. List at least eight human factors-associated problems that can be identified through task analysis.
8. Discuss human factors problem-solving strategies for the following two tasks:
   - Standing for heavy lifting/performing work in one place or in motion.
   - Seated repetitive work with light parts.
9. What are the major elements of OSHA ergonomics guidelines?
10. List at least ten human factors-related issues that must be considered early in a safety program which can directly or indirectly affect safety.

## References


1. Heinrich, H.W., Industrial Accident Prevention, 4th addition, McGraw-Hill, New York, 1959.
2. Chapanis, A., Man-Machine Engineering, Wadsworth Publishing, Belmont, California, 1965.
3. Dhillon, B.S., Engineering Safety: Fundamentals, Techniques, and Applications, World Scientific Publishing, River Edge, New Jersey, 2003.
4. Nation Institute of Occupational Safety and Health (NIOSH), Department of Health and Human Services, Washington, D.C., 2002.
5. Hammer, W., Price, D., Occupational Safety Management and Engineering, Prentice Hall, Upper Saddle River, New Jersey, 2001.
6. Beech, H.R., Burns, L.E., Sheffield, B.F., A Behavioral Approach to the Management Stress, John Wiley and Sons, New York, 1982.
7. Grimaldi, J.V., Simonds, R.H., Safety Management, Richard D. Irwin, Homewood, Illinois, 1989.
8. Warshaw, L.J., Occupational Stress, in Introduction to Occupational Health and Safety, edited by J. Ladou, National Safety Council, Chicago, 1986, pp. 140–150.
9. National Safety Council, "Making the Job Easier: An Ergonomics Idea Book", National Safety Council, Chicago, 1988.
10. Meister, D., Human Factors in Reliability, in Reliability Handbook, edited by W.G. Ireson, McGraw-Hill, New York, 1966, pp. 12.2–12.37.
11. Goetsch, D.L., Occupational Safety and Health, Prentice Hall, Englewood Cliffs, New Jersey, 1996.

12. National Safety Council, "Ergonomics: A Practical Guide", National Safety Council, Chicago, 1988.
13. LaBar, G., Ergonomics: The Mazda Way, Occupational Hazards, April 1990, pp. 44–45.
14. Occupational Safety and Health Administration (OSHA), Report No. OSHA 3123, Department of Labor, Washington, D.C., 1991.
15. System Safety Handbook, Federal Aviation Administration (FAA), Washington, D.C., 2000.

# chapter fourteen

# Software and robot safety

## 14.1 Introduction

Nowadays, computers have become very important element of day-to-day life, and they are made up of both hardware and software components. Today, much more money is spent for developing computer software than hardware. Needless to say, software has become a driving force in the computer industrial sector along with rapidly growing concerns for its safe functioning. In many applications, its proper functioning is so important that a simple malfunction/failure can result in a large-scale loss of human lives. For example, Paris commuter trains serving around 1 million passengers daily very much depend on software signaling [1].

Nowadays, robots are increasingly being used in the industrial sector for performing various types of tasks including materials handling, arc welding, routing, and spot welding. Currently, there are over 1 million robots in use in the industrial sector around the globe. The history of robot safety is traced back to 1980s with the development of the American National Standard for Industrial Robots and Robot Systems: Safety Requirements [2], and the Japanese Industrial Safety and Health Association document entitled 'An Interpretation of the Technical Guidance on Safety Standards in the Use, etc., of Industrial Robots' [3].

This chapter presents various important aspects of software and robot safety.

## 14.2 Software hazard causing ways

There are various ways in which software can cause/contribute to a hazard. Seven of these ways are as follows [4, 5]:

- **Way I:** Provided incorrect solution to a problem.
- **Way II:** Poor timing of response for an adverse situation.
- **Way III:** Performed a function not required.
- **Way IV:** Conducted a function out-of-sequence.
- **Way V:** Failure to recognize a hazardous condition requiring a corrective measure.
- **Way VI:** Poor response to a contingency.
- **Way VII:** Failure to conduct a required function.

## 14.3 *Basic software system safety-related tasks and software quality assurance organization's role in regard to software safety*

Although there are many software system safety-related tasks, some of the basic ones include [6]:

- Establish appropriate safety-related software test plans, test procedures, test descriptions, and test case-related requirements.
- Highlight the elements of the software that control safety–critical operations and then direct all necessary safety analysis and tests on those specific functions and on the safety–critical path that leads to their execution.
- Develop a tracking system within the software framework along with system configuration control structure for assuring the traceability of safety-related requirements and their flow through documentation.
- Clearly show the software system safety-associated constraint consistency in regard to the software requirements specification.
- Develop on the basis of highlighted software system safety-related constraints the system-specific software design criteria and requirements, computer–human interface-related requirements, and testing-related requirements.
- Review with care the test results concerning safety-related issues and trace the highlighted safety-related software problems right back to the system level.
- Conduct any special safety-related analyses, e.g., computer–human interface analysis or software fault tree analysis (SFTA).
- Highlight with care all the safety–critical elements and variables for use by code developers.
- Trace all types of safety-related requirements constraints right up to the code.
- Trace all highlighted system hazards to the hardware–software interface.

Past experiences indicate that generally software safety-related requirements originate from the end-user organization and flows down to the development organization. Thus, the Software Quality Assurance Organization possesses the appropriate ingredients to best conduct the task of highlighting, controlling, documenting, and reducing software safety-associated risk. In this regard, the organization's roles include [7]:

- Define user safety-related requirements, the operational concept, and the operational doctrine.

- Develop the operational safety-related policy that highlights acceptable risks along with operational alternatives to hazardous operations.
- Define requirements for conducting operational safety-related reviews.
- Define safety-related criteria for system acceptance.
- Conducting safety-related reviews and audits of operational systems regularly.
- Investigate, evaluate, resolve, and document all reported safety-associated operational incidents.
- Approve the results of safety-related testing prior to releasing the systems.
- Chair operational safety review panels.

## 14.4 Software safety assurance program

A safety assurance program within the organization's framework basically involves three maturity levels: A, B, and C [7]. Maturity level A is concerned with the development of company culture being clearly aware of software safety-associated issues. More clearly, in this case, all personnel involved with software development work according to standard development rules and apply them consistently.

Maturity level B is concerned with the implementation of a development process involving safety assurance-associated reviews and hazard analysis for highlighting and eliminating safety–critical conditions before being designed into the system. Finally, maturity level C is concerned with the utilization of a design process that documents the results and implements continuous improvement methods for eradicating safety–critical errors out of the system software.

Nonetheless, some the items to be considered during the implementation process of a software safety assurance program include [7, 8]:

- All software system safety-associated requirements are consistent with contract requirements.
- Software system safety is quantifiable to the stated risk level using the usual measuring methods.
- All software system hazards are highlighted, evaluated, tracked, and eradicated as per requirements.
- All human–computer interface-associated requirements are clearly consistent with contract requirements.
- Past software safety-related data are fully considered and utilized in all potential software development projects.

- Software system safety is clearly addressed in terms of a team effort that involves groups such as quality assurance, engineering, and management.
- Software system safety-related requirements are developed and stated as a component of the organization's design policy.
- All changes in design, configuration, or requirements are conducted such that they still maintain an acceptable level of risk.

Finally, it is to be noted that a software safety assurance program also must consider with care factors such as recording of safety-related data, assuring that safety is designed into the system timely and cost-effectively, and minimizing risk when using and accepting items, such as designs, new materials, and test and production methods [7, 8].

## 14.5 Software hazard analysis methods

There are many methods that can be used for performing various types of software hazard analysis. Most of these methods are as follows [8–12]:

- Code walk-through.
- Software sneak circuit analysis.
- Software fault tree analysis.
- Proof of correctness.
- Event tree analysis.
- Software/hardware integrated critical path.
- Monte Carlo simulation.
- Cause-consequence diagrams.
- Hazard and operability studies.
- Design walk-through.
- Cross reference-listing analysis.
- Petri net analysis.
- Failure modes and effect analysis (FMEA).
- Desk checking.
- Structural analysis.
- Nuclear safety cross-check analysis.

The first five of the above 16 methods are described below [8–12].

### 14.5.1 Code walk-through

This is a very useful method for improving safety and quality of software products. The method is basically team effort among professionals such as system safety professionals, software programmers, software engineers, and program managers. Code walk-throughs are in-depth reviews of the software in process as well as inspection of the software functionality. All logic branches as well as each statement's function are discussed with care at a significant length. More clearly, this process provides a good check and balances system of the software developed.

The system reviews the software's functionality and compares it with the stated system requirements. This provides a verification that all stated software safety-related requirements are implemented correctly, in addition to the determination of functionality accuracy. Additional information on this method is available in Ref. [12].

### 14.5.2 *Software sneak circuit analysis*

This method is used for highlighting software logic that causes undesired outputs. More clearly, program source code is converted to topological network trees and the code is modeled by employing six basic patterns: single line, parallel line, trap, iteration loop, entry dome, and return dome. All software modes are modeled employing the basic patterns linked in a network tree flowing right from top to bottom.

The analyst asks questions on the use and interrelations of the instructions considered as the structure's elements. The effective answers to questions asked are very useful in providing clues that highlight sneak conditions (an unwanted event not caused by component failure) that may result in undesirable outputs. The analyst searches for the following four basic software sneaks:

- Existence of an undesired output.
- Wrong timing.
- The undesired inhibit of an output.
- A program message that poorly describes the actual condition.

All the clue-generating questions are taken from the topograph denoting the code segment, and at the discovery of sneaks, the analysts conduct investigative analyses for verifying that the code does not indeed generate the sneaks. Subsequently, all the possible impacts of the sneaks are assessed and required corrective measures recommended.

### 14.5.3 *Software fault tree analysis (SFTA)*

This method is an offshoot of the fault tree analysis (FTA) method developed in the early years of 1960s at the Bell Telephone Laboratories for analyzing the Minuteman Launch Control System from the safety aspect [13]. SFTA is used for analyzing software design safety, and its main objective is to demonstrate that the logic contained in the software design will not cause system safety-related failures, in addition to determining environmental conditions that may lead to the software causing a safety-related failure [14].

SFTA proceeds in a similar manner to hardware FTA described in Chapter 4, and it also highlights software–hardware interfaces. Although fault trees for both software and hardware are developed quite separately, they are linked together at their interfaces for allowing total system analysis. This is very important because it is impossible to develop software safety-related procedures in isolation, but must be considered as a part of the total system safety.

Finally, it is to be noted that although SFTA is a very good hazard analysis method, it is a quite expensive tool to use.

### 14.5.4 *Proof of correctness*

This is a quite useful method for performing software hazard analysis. It decomposes a program under consideration into a number of logical segments, and for each and every segment input/output assertions are defined. Subsequently, the involved software professional conducts verification from the perspective that each and every input assertion and its associated output assertion are true and that, if all of the input assertions are true, then all of the output assertions are also true.

Finally, it is to be noted that this method makes use of mathematical theorem proving concepts for verifying that a given program is clearly consistent with its associated specifications. Additional information on this method is available in Refs. [9–11].

### 14.5.5 *Event tree analysis (ETA)*

This method models the sequence of events resulting from a single initiating event, and in regard to its application to software, the initiating event is taken from a code segment considered safety critical (i.e., suspected of error or code inefficiencies). Normally, ETA assumes that each and every sequence event is either a success or a failure.

Some of the important factors associated with ETA are as follows [15]:

- It is a very important tool for identifying undesirable events that need further investigation using the FTA method.
- Generally, the method is used for conducting analysis of more complex systems than the ones handled by the FMEA method.
- It is difficult for incorporating delayed success or recovery events.
- It always leaves some room for missing important initiating events.

Additional information on this method is available in Refs. [15, 16].

## 14.6 Robot safety problems and accident types

Over the years, various safety professionals have studied robot safety-related problems and have highlighted many unique robot safety-related problems. As per Ref. [17], some of these problems are as follows:

- A robot may lead to a high risk of fire or an explosion if it is installed in unsuitable environments.
- Robots create potentially hazardous conditions because they frequently manipulate objects of different sizes and weights.
- Robots may go out of their programmed zones and strike something or throw objects if a hydraulic, mechanical, or control failure occurs.
- Generally, humans give great attention to a robot's presence, and they are often quite ignorant of the potential related hazards.
- Generally, robots function closely with other machines and humans under the same environments. More clearly, humans are subject to collisions with moving robot parts, tripping over loose control/power cables (if any), etc.
- Robot mechanical design-associated problems directly or indirectly might cause hazards such as pining, pinching, and grabbing.
- Safety-associated electrical design-related problems such as potential electric shock, fire-related hazards, and poorly designed power sources can occur in robots.
- Robot maintenance-related procedures may result in hazardous situations.

Although there are many different types of robot accidents, they can be categorized under four broad classifications, as shown in Fig. 14.1 [18].

The collision/impact accidents are concerned with situations where unpredicted component-related failures, unpredicted movements, or unpredicted program changes related to the arm of robot or peripheral equipment lead to contact accidents. The trapping/crushing accidents are concerned with situations where a person's limb or other body part is trapped between a robot's arm and other peripheral equipment or the person is physically driven into and crushed by other peripheral equipment of various types.

The mechanical part accidents include situations where the breakdown of the robot's drive-related parts/components, power source, peripheral equipment, or its tooling or end-effectors occurs. Some examples of mechanical failures are the failure of end-effect or power tools, failure of gripper mechanism, and release of parts. The miscellaneous accidents include all those accidents that cannot be categorized under the other three classifications. More simply, they refer to all other accidents that occur from working with robots. For example, environmental accidents

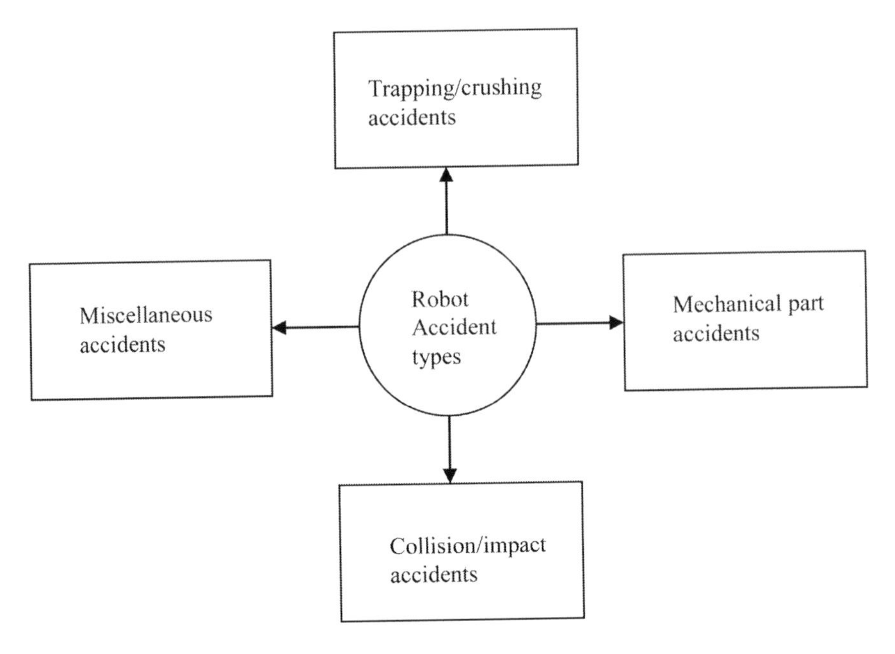


***Figure 14.1*** Robot accident classifications.

from dust, metal spatter, arc flash, electromagnetic or radio-frequency interference, dangerous high-pressure cutting streams or whipping hose-related hazards because of ruptured hydraulic lines, and tripping hazards from equipment and power cables on the floor.

## *14.7 Robot hazard causes*

There are many causes of robot hazards. The main ones of these causes are shown in Fig. 14.2 [18, 19]. Control errors are an important cause of robot hazards, and they can occur due to various reasons including faults in the pneumatic, electrical, or hydraulic controls associated with the robot system. Power systems are another important cause of robot hazards because pneumatic, electrical, or hydraulic power sources that have malfunctioning transmission or control parts/components in the robot power system can cause a disruption in electrical signals to power control/supply lines.

Environmental sources are also an important cause of robot hazards. For example, radio frequency or electro-magnetic interference (transient signals) can exert an undesirable influence on robot operation and significantly increase the probability of injury to persons performing tasks in the area. The unauthorized entry into the safeguarded robot zone is quite hazardous because the individual concerned may be unfamiliar with the activation status or the safeguards in place.

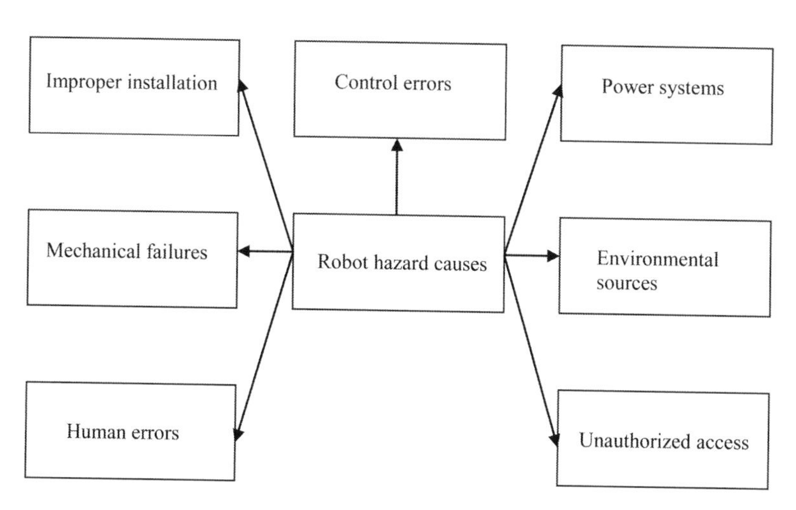


***Figure 14.2*** Robot hazards' main causes.

Human errors are another important cause of robot hazards, and a quite common human error associated with robots is the wrong activation of the 'teach pendant' or control panel. Some other examples of human errors are interfacing already activated peripheral equipment, placing oneself in a quite hazardous position while programming the robot, and connecting live input–output sensors to the microprocessor or a peripheral. Mechanical failures also can result in robot hazards because functioning programs may not account for cumulative part/component failure, thus making it possible for the occurrence of unexpected or faulty operation.

Finally, improper installation can also cause various types of robot hazards. More clearly, the equipment layout, requirement design, utilities, and a robot system's facilities, if executed incorrectly, can lead to inherent hazards.

## 14.8 *Safety considerations in robot life cycle*

Robot life cycle may be divided into four phases, as shown in Fig. 14.3 [8].

Each phase shown in Fig. 14.3 is described below [3, 20–23].

### 14.8.1 *Design phase*

In this phase, the safety considerations may be classified under the following three categories:

- **Category I: Mechanical:** This category includes safety considerations such as designing teach pendant ergonomically; eliminating sharp corners; putting guard belts on items such as gears, pulleys,

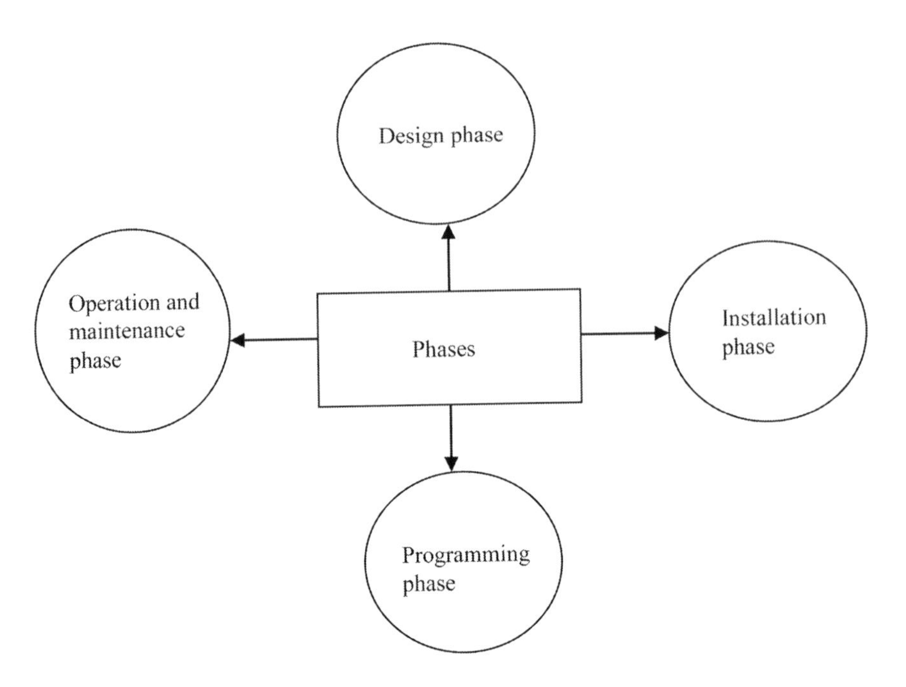


*Figure 14.3* Robot life cycle phases.

and belts; having mechanisms for releasing the stopped energy; and having several emergency stop buttons.

- **Category II: Electrical:** This category includes safety considerations such as minimizing the effects of electromagnetic and radio frequency interferences, having a fuse 'blow' long before human crushing pressure is experienced, designing wire circuitry capable of stopping the robot's movement and locking its brakes, and having built-in hose and cable routes using adequate insulation, sectionalization, and panel covers.
- **Category III: Software:** This category includes safety considerations such as prohibiting a restart by merely resetting a single switch, examining periodically the built-in self-checking software for safety, having a standby power source for robots operating with programs in random access memory, providing a robot motion simulator, having built-in safety commands, and using a procedure or checks to determine why a failure occurred.

### 14.8.2 *Installation phase*

There are many installation phase robot safety-related considerations. Some of these are as follows:

- Placing robot controls outside the hazard zone.
- Providing adequate illumination to personnel concerned with robots.
- Installing the necessary interlocks to interrupt robot motion.
- Ensuring the visibility and accessibility of emergency stops.
- Using vibration-reducing pads when appropriate.
- Installing electrical cables according to electrical codes.
- Placing an appropriate shield between the robot and personnel.
- Distancing circuit boards from electromagnetic fields.
- Indentifying the danger zones with the aid of signs, line markings, code, etc.
- Installing interlocks and sensing devices.

### *14.8.3 Programming phase*

There are also many programming phase safety-related considerations for robots. Some these are as follows:

- Marking the programming positions.
- Adjustable limits for robot axes.
- Pressure sensitive mats on the floor at the position of the programmer.
- A manual programming device containing an emergency off switch.
- Hold-to-run buttons.
- Mandatory reduced speed.

### *14.8.4 Operation and maintenance phase*

This is a very important phase of the robot life cycle, and many robot hazards are confined to this phase. Some of the safety measures associated with this phase are as follows:

- Developing appropriate safety operations and maintenance procedures.
- Blocking out all associated power sources during the maintenance activity.
- Minimizing the risk of fire by using non-flammable liquids for lubrication and hydraulics.
- Keeping (if possible) at least one extra person in the vicinity during the robot repair process.
- Minimizing the potential energy of an unexpected motion by having the robot arm extended to its maximum reach as much as possible.
- Ensuring the operational readiness of all associated safety devices.
- Ensuring that only authorized and trained individuals operate and maintain robots.

- Providing appropriate protective gear to all involved individuals.
- Conducting preventive maintenance regularly and using only the approved spare parts/components.
- Ensuring the proper functionality of all emergency stops.
- Posting the operating weight capacity of the robot.
- Investigating any fault or unusual robot motions immediately.

## 14.9 Robot safeguard approaches

There are many robot safeguard approaches. Six of these approaches are as follows [20, 24]:

- **Flashing lights.** These lights are installed at the perimeter of the robot-working zone or on the robot itself for alerting people that robot programmed motion is happening or could happen any moment.
- **Intelligent systems.** These systems use intelligent control systems for safeguarding, and they use avenues, such as hardware, software, and sensing, in making decisions.
- **Physical barriers.** The basic objective of these barriers is to stop humans from reaching over, around, under, or through the barrier into a robot's forbidden work zone. Some examples of physical barriers are safety rails, tagged rope barriers, chain link fences, and plastic safety chains.
- **Warning signs.** These signs are used in environments where robots, by virtue of their speed, inability, and size for imparting excessive force, cannot injure humans. Nonetheless, past experiences over the years clearly indicate that the warning signs are extremely useful for all types of robot environments.
- **Electronic devices.** These devices are basically the application of ultrasonic for perimeter control to have protection from intrusion. The perimeter control electronic barriers use active sensors for detecting intrusions.
- **Infrared light arrays.** The commonly used linear arrays of infrared sources are light curtains. Although, the past experiences over the years indicate that the light curtains are quite reliable, from time to time false triggering may take place due to factors such as flashing lights, heavy dust, or smoke because of misalignment of system parts.

## 14.10 Problems

1. Write an essay on software and robot safety.
2. List at least seven software hazard causing ways.

3. List at least eight basic software system safety-related tasks.
4. Describe the software safety assurance program.
5. Describe the following software hazard analysis methods:
   - Proof of correctness.
   - SFTA.
6. List at least seven unique robot safety-related problems.
7. What are the main causes of robot hazards?
8. Discuss safety considerations in robot life cycle.
9. Describe at least five robot safeguard approaches.
10. Describe the following software hazard analysis methods:
    - Code walk-through.
    - Software sneak circuit analysis.

## *References*

1. Cha, S.S., Management Aspects of Software Safety, Proceedings of the 8th Annual Conference on Computer Assurance, 1993, pp. 35–40.
2. ANSI/RIA R15.06-1986, American National Standard for Industrial Robots: Safety Requirements, American National Standards Institute (ANSI), New York, 1986.
3. Japanese Industrial Safety and Health Association, An Interpretation of the Technical Guidance on Safety Standards in the Use, etc., of Industrial Robots, Japanese Industrial Safety and Health Association, 5-35-1, Shiba, Minato-ku, Tokyo, 1985.
4. Leveson, N.G., Software Safety: Way, What, and How, Computing Surveys, Vol. 18, No. 2, 1986, pp. 125–163.
5. Friedman, M.A., Voas, J.M., Software Assessment, John Wiley and Sons, New York, 1995.
6. Leveson, N.G., Software, Addison-Wesley Publishing, Reading, Massachusetts, 1995.
7. Mendis, K., Software Safety and Its Relation to Software Quality Assurance, in Handbook of Software Quality Assurance, edited by G.G. Schulmeyer, J.I. McManus, Prentice Hall, Upper Saddle River, New Jersey, 1999, pp. 669–679.
8. Dhillon, B.S., Engineering Safety: Fundamentals, Techniques, and Applications, World Scientific Publishing, River Edge, New Jersey, 2003.
9. Hansen, M.D., Survey of Available Software-Safety Analysis Techniques, Proceedings of the Annual Reliability and Maintainability Symposium, 1989, pp. 46–49.
10. Ippolito, L.M., Wallace, D.R., A Study on Hazard Analysis in High Integrity Software Standards and Guidelines, Report No. NISTIR 5589, National Institute of Standards and Technology, U.S. Department of Commerce, Washington, D.C., January, 1995.
11. Hammer, W., Price, D., Occupational Safety Management and Engineering, Prentice Hall, Upper Saddle River, New Jersey, 2001.
12. Sheriff, Y.S., Software Safety Analysis: The Characteristics of Efficient Technical Walk-Throughs, Microelectronics and Reliability, Vol. 32, No. 3, 1992, pp. 407–414.

13. Dhillon, B.S., Singh, C., Engineering Reliability: New Techniques and Applications, John Wiley and Sons, New York, 1981.
14. Leveson, N.G., Harvey, P.R., Analyzing Software Safety, IEE Transactions on Software Engineering, Vol. 9, No. 5, 1983, pp. 569–579.
15. Dhillon, B.S., Design Reliability: Fundamentals and Applications, CRC Press, Boca Raton, Florida, 1999.
16. Cox, S.J., Tait, N.R.S., Reliability, Safety, and Risk Management, Butterworth-Heinemann, London, 1991.
17. Van Deest, R., Robot Safety: A Potential Crisis, Professional Safety, January 1984, pp. 40–42.
18. "Industrial Robots and Robot System Safety", Chapter 4, in OSHA Technical Manual, Occupational Safety and Health Administration (OSHA), Department of Labor, Washington, D. C., 2001.
19. Ziskovsky, J.B., Working Safely with Industrial Robots, Plant Engineering, May 1984, pp. 81–85.
20. Dhllion, B.S., Robot Reliability and Safety, Springer-Verlag, New York, 1991.
21. Russell, J.W., Robot Safety Considerations: A Checklist, Professional Safety, December 1983, pp. 36–37.
22. Blache, K.M., Industrial Practices for Robotic Safety, in Safety, Reliability, and Human Factors, edited by J.H., Graham, Van Nostrand Reinhold, New York, 1991, pp. 34–65.
23. Nicolaisen, P., Ways of Improving Industrial Safety for the Programming of Industrial Robots, Proceedings of the 3rd International Conference on Human Factors in Manufacturing, November 1986, pp. 263–276.
24. Fox, D., Robotic Safety, Robotic World, January/February, 1999, pp. 26–29.

# Index